PARABLES
Architecture with Hidden Secrets

寓言书　建筑的秘密

　　　　　温子先　博士的建筑实践

Architecture of **Andy Wen** Ph.D
(2008-2014)

\+　　　　\+

Aedas

\+　　　　\+

温子先　编著

天津大学出版社　
TIANJIN UNIVERSITY PRESS

Preface

Preface

Architecture Wants to SPEAK

Persistent in his architectural design practice, Andy Wen continuously focuses on the subject of "Ambiguity in Collective Representation" which investigates the representational ability of architecture.

In recent history, architecture has been quiet. Beginning at the end of the last century, waves of philosophical criticism challenged the Modernist International Style that ignored both nature and the city. Architects produced expressionless buildings unable to satisfy the public's architectural desires in the post industrial period. In today's "mediatized" society, architecture cannot remain silent; architecture wants to, and must speak.

How Architecture SPEAKS

Architecture does not speak in a literal manner. Architecture communicates through its ability to provoke the senses. It also communicates with abstract metaphors and uses methods that break down cultural barriers, similar to a painting or a sculpture. That which provokes the senses is typically "ambiguous" rather than direct interpretations.

The intent architecture communicates often results in a "collective" meaning.

Messages can often have multiple depths of meaning as they are interpreted by people of different culture, race, religion etc. A different view can often be the spark of a new creation and translated into architecture.

The purpose of architecture today challenges these basic dimensions of "Ambiguity in Collective Representation."

What Is a PARABLE

Parable is "the designed use of language purposely intended to convey a hidden and secret meaning other than that contained in the words themselves, which may or may not bear a special reference to the listener or reader."

A parable is a short tale that illustrates universal truth, one of the simplest of narratives.

A parable is like a metaphor that has been extended to form a brief and coherent fiction.

Parables are favored in the expression of spiritual concepts.

Architecture and PARABLES

This is a story represented by architecture.

This is both an architectural portfolio and a book of parables.

Architecture here does not mean a "practical, stable or beautiful" entity. It is a meaningful living entity which breathes, talks and waits to be experienced and appreciated.

You are invited to join the following thinking processes about the city, about the future, about all of us.

Preface

建筑想要说话

温子先先生在他的建筑设计实践中,一直在探讨"综合再现的不确定性"。

该课题的关键在于建筑的表意和叙事能力。

20世纪末掀起的哲学思潮对现代主义国际化风格的建筑进行了全面的批判,这些建筑往往无法融入自然及城市之中,千篇一律的单调构筑物显然已无法满足后工业时代公众对建筑的诉求。在当下的信息社会中,一切都被媒体化,建筑也不能再保持沉默,建筑也需要开口说话了。

建筑如何说话

建筑超越构筑物的范畴,在于建筑拥有激发感知的能力。它可以跨越不同文化之间的障碍,传达多重意义的隐喻,如同伟大的绘画和雕塑作品。在这里,被感知的是一种"不确定性"("多义模糊")的而非直接单一的阐释。

通过建筑为媒介所实现的沟通常以"综合性"范畴为结果,传达的讯息被拥有不同文化、种族、宗教信仰的解读者翻译为不同深度和层面的意义。这些不同的观点能激发出新的创作灵感,并转变为富有意义的设计表达。

当今建筑学的目的,在一定程度上,考验着"综合再现的不确定性"的这些基本维度。

什么是寓言

"寓言"是"一种语言的使用方式,旨在传达被隐藏在字面背后的神秘意义,而且同一个故事对于不同的聆听者和阅读者并不只具有单一的理解方式"。

寓言是一个简短的故事,却以一种最简单的叙述方式阐述了普遍的真理。

寓言就如同是一个隐喻,由此可以扩展为简要连贯的虚构故事。

寓言热衷于传达某种精神理念。

建筑与寓言书

将在您面前呈现的是一个由建筑所叙说的故事。

这是一本建筑作品集,也是一本寓言书。

这里的建筑不只是"实用、坚固、美观"的无机体,它们是有意识、有生命的有机体,它们会呼吸、会诉说,并随时恭候您的观赏和品鉴。

邀请您加入这个思考的过程:关于城市,关于未来,关于我们……

About the Preface

In this book there are simple conceptual design descriptions at the beginning of each design. The function of this writing is equivalent to contemporary artwork explanations, used to articulate design intent and requiring viewers to experience the metaphors within. These descriptions are a guide offering a basic explanation and perhaps more important inspiration on multiple levels of intellectual understanding.

During the reading process, one should carry on a dialogue with the work; listen and experience the joy and stimulation of ambiguity in this collection of representation. Architecture surely can communicate more than words.

About the Incomplete

As a general statement, all the projects described within this book are incomplete. The following are three perspectives on architectural incompletion.

Incomplete design: The architectural design process is generally divided into phases beginning with conceptual design and then moving through schematic design, design development, and construction documentation. These phases are typically bookended by a project feasibility study prior to actual design and culminate with the actual building construction, when the architecture is presented for all to view. Many of the cited projects are evolving within different phases of design; therefore, the projects are in an incomplete state.

Incomplete building: Even a built project as it exists is incomplete by definition. Since the collapse of classical physics theory, people have recognized that the concept of absolute order cannot explain our current universe. It is the same situation in architecture; our environment is changing; the user is changing; weather and time are wearing out the built environment. Therefore, facing a rapidly developing world, we tolerate visual chaos where complications amplify ambiguity. The designated point of architectural completion has become relatively vague. When changing becomes the normal state, incompletion seems to accompany the entire practice of architectural design.

Incomplete dialogue: Revisiting architecture's expressive role, one finds it can be similar to contemporary art. The relationship between architecture and the viewer is not a singular imperative instruction, but is rather a dialogue. Architects give meaning to the building establishing the first step of the dialogue; the response and thus the completion of the dialogue is in the hands of the spectator. Each observer interprets and gives meaning to the architecture. One thousand viewers will generate one thousand interpretations, which provides the fascination of "ambiguity."

关于每个项目的引言

书中的每个项目都有一段引言，对设计概念和建筑所表达的内容进行简单的描述。项目引言的功能接近于当代艺术作品的说明，它并不能穷尽作品所要阐述的意义，更多的含义在文字之外，需要观赏者的悉心体会。项目引言更像是一种引导，对作品作了最基本的解释，目的是引发更多层面的解读和思考。

读者在阅读本书的过程中，尝试用心来和设计作品对话，聆听和感受那种综合再现的不确定性所带来的乐趣和启发。建筑比起文字可以传递的内容更多、更丰富。

关于建筑的未完成

可以说本书中的作品都是"未完成"的，这里就三个层面来解释建筑的"未完成"。

设计的"未完成"：建筑设计一般分为概念设计、方案设计、节点设计、施工图设计，前期可行性研究工作，以及后期根据实际设计进行的施工过程，一栋建筑物才能以实体的方式呈现于我们面前。书中的多数项目，还处于设计过程中的不同阶段，因此这些项目也还处于"未完成"的状态。

建筑的"未完成"：即使是已建成的项目，也是某种意义上的"未完成"。从经典物理学的崩溃开始，人类已经意识到绝对的静态秩序观已经不能解释我们所处的宇宙了。对于建筑也是这样，使用者在变、周围的环境在变、时间和气候对建筑物也在不断地改变，面对一个飞速变化的世界，我们忍受着视觉的混沌，其复杂性增加了不确定性。建筑"完成"的那个时间点也被模糊了，变化成为一种常态，"未完成"似乎伴随建筑的整个生命历程。

对话的"未完成"：回到建筑的表意功能，建筑与观看者之间并不是单向的布道关系，而是一种双向的对话关系，这有些类似于当代艺术作品与观赏者的关系。建筑师将自己要表达的内容赋予建筑的时候只完成了对话的第一步，对建筑的反馈和真正完成的时间点掌握在观赏者手中。建筑所传达的意义也由每一个观赏者来解读，一千个人有一千种不同的理解，这是"不确定性"的魅力所在。

Contents

The Organizational Logic of This Book

The main body of this work is divided into five chapters. The first chapter focuses on Andy's design philosophy. Chapter Two, Three and Four collect his tangible projects with Aedas. Chapter Five focuses on his personal growth and social background. PARABLES interprets architecture as a form of narration. The projects are subcategorized into tales of inanimate objects, tales of creatures and tales of humanity.

Contents

009

1   Ambiguity in Collective Re-presentation _ 012

2   Tales of Inanimate Objects
Scholar Stone _
Xi'an Jiaotong-Liverpool University Administration and Information Building _ 024
Lucky Cloud _
Xuzhou Suning Plaza _ 036
Landscape _
333 Shunjiang Road Mixed-use Project _ 048
River Pebble _
Taipei Nangang Office Tower _ 062
Jade _
Shanghai EXPO Urban Best Practice Area E06-04 Block _ 070
River _
Tianjin Century Times Art Museum _ 078
Alluvial Plain _
Cheng Ying Center _ 086

3   Tales of Creatures
Hibiscus _
Chengdu Jinniu District North 1st Ring Rd. N Project _ 96
Oasis_
Xiamen Green Land Mixed-use Project _104
Tea Plantation_
Suzhou Hong Leong City Center _ 116
Butterfly _
Sino-Singapore Tianjin Eco-city Plot 8 &17 Planning _ 126
Dragon 1 _
Nanjing Hunan Road Plaza _ 136
Dragon 2 _
Hengqin International Financial Center _ 146

4   Tales of Humanity
African Dance _
Angola Mixed-use Commercial Center _ 162
Lover_
Nanjing Hexi Olympic Plaza _ 170
City Files I — Modern Church _
Huai'an Suning Plaza _ 176
City Files II — History _
COFCO Shanghai Joy City Phase 2 _ 182
City Files III — Hakka Walled Village _
Wuxi Metro Mixed-use Project _ 194
Gift Box _
Beijing Suning Plaza _ 206
Ink Painting _
Beijing Artist Village Gallery _ 214
The Kiss _
Haikou Changliu West Coast President Tower _ 228
The Staircase _
Keppel Land Chongqing Mixed-use Development Concept Masterplan _ 234
Memory _
Guangzhou Nansha District Jiaomen River Central Area Southern Waterfront _ 246

5   Background of Andy _ 259
Social Background of Andy's Design _ 261

关于本书的顺序编排

全书分为五个部分,第一部分主要介绍温子先博士建筑创作背后的设计思想。第二到第四部分收录了温先生在凯达环球期间主持设计的具体项目。第五部分为温先生个人经历和社会背景。《寓言书》尝试探讨建筑的叙事性,项目编排借鉴了寓言故事的分类方式,依照设计概念归类为无机物篇、生物篇和人类篇。

Contents

1
综合再现的不确定性 _ 012

2 无机物的故事
太湖石 _
西交利物浦大学行政信息楼 _ 024
祥云 _
徐州苏宁广场 _ 036
山水 _
门里成都顺江路333号综合体项目 _ 048
鹅卵石 _
台北南港办公大楼 _ 062
玉 _
上海世博会城市最佳实践区E06-04地块 _ 070
河 _
天津坝上春秋美术馆 _ 078
冲积平原 _
诚盈中心 _ 086

3 生物的故事
芙蓉花开 _
成都金牛区北一环项目 _ 096
绿洲 _
厦门绿地综合体项目 _ 104
茶园 _
苏州丰隆城市中心 _ 116
蝴蝶 _
中新天津生态城开发区第8、17地块规划方案 _ 126
盘龙 _
南京湖南路广场 _ 136
蛟龙 _
横琴国际金融中心大厦 _ 146

4 人类的故事
非洲舞 _
安哥拉综合商业中心 _ 162
情人 _
南京河西奥体新城广场 _ 170
城市档案一——都市教堂 _
淮安苏宁电器广场 _ 176
城市档案二——历史 _
上海大悦城项目二期 _ 182
城市档案三——土楼 _
无锡地铁综合体项目 _ 194
礼品盒 _
北京苏宁电器广场 _ 206
水墨画 _
北京世纪国际艺术城美术馆 _ 214
吻 _
海口长流西岸首府 _ 228
阶梯 _
吉宝置业重庆凯旋路项目概念规划 _ 234
记忆 _
广州南沙新区蕉门河中心南部滨水角 _ 246

5
温子先先生的经历 _ 259
温子先先生设计创作的社会背景 _ 261

CHAPTER ONE

Ambiguity in Collective Re-presentation

There is a dilemma on how to reflect the change of time and with the same effort return to the basics. This is the focal point of modern architecture in both China and the Western World. The questions about contemporary architectural existence involve borders, society, culture, arts and crafts. We live now in a complex and intricate world; how to expand knowledge beyond one's borders and remain compassionate about social issues are now definitely major lessons for architects.

In response to the evolutionary transformation of metropolitan lifestyle and daily routine, early modernist architecture focused on the resolution of functional issues. Meanwhile, the modernist movement overlooked other important responsibilities such as understanding cultural and structural frameworks, while maintaining a correspondence with human needs.

Addressing the question "how do we create communication within today's cultural framework?", Andy Wen recognizes "It is time to re-start the communication process for and within today's architecture" and has proposed the concept of "Ambiguity in Collective Re-presentation."

Here are three approaches to evaluate and understand the concept:

"Collective" responds to today's media based society. Living in compressed time, people are constantly forced to face the feeling of loss of space and are overwhelmed with information from all directions and time periods. The world is becoming both monotonous and complicated simultaneously. A healthy architecture must be able to express the denotation in collective ways and communicate at many complex social levels. Beyond the value of aesthetics, the most important is the capacity to interpret what is at hand and satisfy the needs of the society.

"Representation" responds to the aspiration of returning to fundamentals. The idea of a "common relationship" explains the concept that every existence in the universe is in some degree associated with another. In nature, there is never a "new" creation — it is always derived from another object that already exists. Here we are questioning the idea of simple replication of existing objects. We are advocating the idea of "Revolutionary Renewals." Representation is not an empty creation. Representation is a deliberate thinking process about nature, history, culture, race, religion, etc. that inspires new design ideas to transform into a meaningful design expression.

"Ambiguity" responds to an increasingly complicated and chaotic society. With the collapse of the classical world order and the decline of utopia, it is almost impossible to describe our surroundings clearly. The disarray obstructs our sensory world with new obstacles; the world in front of us is hazy. Vague is not a negative word anymore, in some ways it contains a high level of uncertainty that provides us with liberty. Because of this liberty the world is no longer limited by the description of academic authorities. Everyone is free to define his or her own world with an equality that never before existed in structured societies. In architecture, the expression of "ambiguity" is a design metaphor and is always an attractive way of social communication.

Architecture can be described as a building with the ability to self express. Architecture is not a straight forward description. Architecture is more like a parable with numerous metaphors, waiting for new interpretations to give it unexpected meaning time after time.

Modern architecture must break through the contradictions and complications to seek out answers. The following paired concepts outline the search process.

Ambiguity in Collective Re-presentation

第一章

综合再现的不确定性

中国以及西方关于当代建筑的焦点问题即"如何既反映时代,又回归本源",这是一件让人左右为难的事。当代建筑中所存在的问题,涉及边界、社会、文化以及工艺等各个领域。对建筑师而言,由于当今世界比以往更加错综复杂,具备更加广阔的知识面以及对社会的关注变得越来越重要了。

早期的现代主义建筑致力于解决建筑功能的课题,以应对工业革命所带来的城市人类生活方式的剧烈变革。同时,它却忽视了建筑的另一个重要责任:在深层的文化结构上与人们的需求相对应。

如何达到这个文化结构上的对应,温子先认为"对于当代的建筑而言,其与外部世界的交流沟通已迫在眉睫"。他同时提出了"综合再现的不确定性"这一概念。

通过以下三个视角对这个概念进行诠释。

"综合性"对应于信息时代的社会背景。身处被压缩时空中的现代人时时处于空间的丧失中,被来自无时不在、无处不有的讯息所包围,世界由此变得索然寡味却又错综复杂。一栋完整的建筑必须能体现不同层面的内涵,并能与社会的多个层面进行"对话",除了具有美学价值外,更重要的是体现即将来临的趋势以及满足社会需求的能力。

"再现性"对应于回归本源的需求。"普遍联系"理念说明宇宙中存在的每一事物都在某种程度上与其他事物有关联,自然界中从没有一种所谓"新"事物,它总是来源于某些已存在的东西。这里所要质疑的是对已有事物的简单"复制",而提倡一种"革命性的重塑"。"再现"不是一种凭空创造,它是对自然、历史、文化、种族、宗教等的深思熟虑,这一思考过程会激发出新的创作灵感,并将之转变为富有意义的设计表达。

"不确定性"对应于日益复杂和混沌的社会。经典世界的轰然倒塌,乌托邦的黄粱一梦,我们已经很难对周遭进行清晰直接的描述。伴随着弥漫感官的混沌和前所未有的变幻速度,世界在我们眼前面目模糊。混沌早已不是贬义词,某种意义上它所具有的高度不确定性给予人类极大的自由。自由让世界不再只是学术权威口中所描述的那种形象,每个人都可以定义自己的世界,这是人类在结构化社会中从未有过的平等。在建筑中,这种"不确定性"表现为设计意义上的隐喻,也一直是一种非常有吸引力的社会沟通方式。

建筑是能够自我阐述的构筑物。它说的不是直白的表述,而更类似于充满言外之意的寓言故事,等待被解读者一次次赋予新的意义。

当代建筑必须在矛盾性和复杂性中寻求解决方案,下面这些成对的概念勾勒出了这种探索的过程。

Chapter One

Globalization vs. Localization

A sarcastic joke was once told, portraying the United States as having evolved from barbarism to lacking enthusiasm for cultural civilization. Perhaps this is an extreme illustration of cultural colonization, but by definition it is one of the saddest tragedies of civilization. This type of cultural invasion existed not only in the past but is continuously happening in various locations around the world. When cultural entities of different strengths meet they require time to breathe and reflect in order to retain a sense of identity.

Beginning in the 1980s, Western architectural influence and stimulation became visible throughout China. Meanwhile, beneath the great waves of globalization and modernization, the longing for traditional proclamations is gradually increasing in an unprecedented magnitude. The architect is responsible for demonstrating and carrying on local traditions and not simply pursuing the trends that allow the long rooted history of ancient civilizations to vanish.

Foreign vs. Familiar

Asked how to flourish in a local culture within a globalized environment, the architect Renzo Piano pointed out an interesting solution; he proposed the idea of "foreignization" and "familiarization" He believes "foreignization" stimulates awareness and minimizes consumption.

A local architect may have an advantage with his or her knowledge of the local cultural background, but, situated in a familiar background for too long can lead to duplication, simplification and manipulation of historical symbols. Therefore, the architect's mission is to use the "foreign" view to critique unsophisticated kitsch.

Regarding the foreign architect already with a "foreignized" cultural background, the learning is focused on how to "familiarize" with local traditions.

The tension between "foreign" and "familiar" is needed to achieve a "collective representation."

Avant-garde and Kitsch

Kitsch is the reflection in miniature of all types of ridiculous contemporary objects. These objects only consume precious time and money without responding to any fundamental human need. The goal of kitsch is to erase the distinction of ethnic treasures around the world, and without limit of culture or location.

The raison raison d'être of the avant-garde is to surmount the typical inertia and continually move forward. It is the path needed for constant self evaluation and redefinition. No idling or remaining unchanged, the ultimate rationale is to identify a clear path to allow culture to continuously develop in the midst of the chaotic and violent world. Avant-garde predecessors have always been singled out from the mainstream trends; they are the real heroes searching for the sublime.

In modern China, the popularly adopted Neoclassical style among the real estate market is the best example of kitsch that is placing vigilant architects in agonizing states.

Revolution and Evolution

"Revolution" is one of the ingredients needed to achieve in architectural work. Architecture is achieved through the creative effort to elevate a building to architectural status. This requires additional innovative hard work rather than simply imitating or copying others' works. Architecture learns through precedents, and through the dissent that exceeds precedents. Using the inspiration of the "rebellion", architecture continues living and proceeding towards "perfection" as its absolute goal.

Looking back on history, in almost all circumstances, it is only revolutionary achievements that have left a deep mark on history. Moderate evolution has been quickly forgotten. The Parthenon, Pantheon, and the Eiffel Tower are great examples of revolutionary achievement with distinctive individuality. They have become historical revolutionary icons.

For the ingenious architect, "revolution" is a key aspect of the creative work. This is not to suggest rejecting the importance of historical material, as referencing the past is not necessarily copying. The modern architect should learn lessons from previous works and reflect upon how to restore the meaning and energy from the past. The object is to figure out how to link together the abstract metaphors from the modern commercialized society, pop culture, fashion trend, media communication and technology. We wish we have had more passion in relation to the creations of our ancestors and less copying. As it was once said: "Greek culture is extraordinary because it never replicated its ancestors."

Symbol and Language

Architects from the previous centuries studied in depth how architecture could be self explanatory. Charles Moore said "Architecture is a performance art." Richard Meier talked about how people should be able to read a building in the same way as reading a book.

Arata Isozaki once explained how he implemented symbol and meaning: "When I worked in Japan, I used a traditional Japanese sensibility to re-interpret Western form and subject matter. When I was working outside of Japan I then attempted to follow Western logic to analyze Japanese-style architecture."

With continuous progress today's architecture could be symbolic or provide metaphors, yet it does neither. Therefore architects can freely interpret architecture through different perspectives and utilize different kinds of architectural language for their creative work. Karsten Harries has stated: "As long as you are willing to articulate the precise feeling, there is no need to limit the selection of language." Today we need to focus on what to say rather than how to say it.

Metaphor and Sign

"Like a lavish indulgent dream, they come into view in our mind and then settle on a potential association amongst themselves." This description of how metaphors work, approaches a common rhythm of communication: The more apparent the metaphor, the more dramatic the effect. The less a sign is understood, the more attractive and mysterious the sign will be. Just as every Shakespeare devotee understands, an uncertain metaphor intensifies its power.

In architecture, just like an attempt to analyze a joke, the effort to explain a metaphor erases it. Any object that obtains aesthetic quality not only offers itself but is a cue or a mark that reminds us of reality, quietly conveying hidden messages.

Architecture must be diverse not simply transparent. Ambiguity and transparency must exist simultaneously. Architecture is just like an article that can be read to encourage inspiration and reminiscing.

Inspirational thinking is a functional process in only one direction; through this process a series of interesting questions are raised. Even though those questions are vague and unclear, chaotic unconscious relations are valuable when compared with balanced and restrictive thinking. This is very important because it allows for an open ended interpretation. Suggestions and audience reactions give an object both complex and general definition.

Object and Intellectual

For a very long period, architecture seemed merely a place for the human body to live. The function of being a "spiritual dwelling" was totally ignored until the later portion of last century when the idea was again mentioned. As Renzo Piano pointed out: "We have to expand technology's functional capacity till it reaches the psychological awareness level, for only then will it transform the building into a human's architecture."

When processing architectural psychological elements we have to make the reasonable assumption that some symbolic content will be recognized as a starting point. The symbolic meaning here refers to the subjects reflecting history and culture's limitations. Perhaps the most important influence from modern art's predecessor Duchamp was that at one level he clearly articulated the object's identity and at the same time suggested this perception was not the only dimension of interpretation.

The key point of art is to translate visual stimulation into "inspiration for the brain." Architecture shares the same quality

Chapter One

through its "objects" to inspire intellectual synchronization.

Society, Humanity and Nature

The relationship between architecture with society, and humanity with nature has varied throughout different historical periods. Sometimes architecture has harmonized with nature satisfying the needs of users. Sometimes architecture has symbolized power in society. Revisiting architectural history it is not hard to find that the outstanding projects are the ones that have shared all these qualities simultaneously, and successfully managed these relationships with the architecture itself.

Society is changing, people are changing and nature is changing, but the architectural mission remains the same: to keep in balance those three relationships.

Architecture is responsible for humanity's quality of living. Many new building technologies provide more comfortable, healthy micro environments, and this is the sincere meaning of "architecture based on human needs." Architecture can be described as one kind of art. Architecture involves more and more political and social issues that are becoming unavoidable subjects for architects. There is no true utopia; all architecture is corporeal.

Architecture and City

The city is an always evolving illustration and every city contains its own chaotic ethics. This is how a city can honestly express our living environment, and it is why a city is an exhilarating and important artistic framework — a unified art composition.

The city is described as a comprehensive society structure having an attitude and experience with authority. These ideas are parallel to evaluations about culture and nature. The city is often perceived as an approved single purpose architectural entity. As an architectural hypothesis this is worthless. The best architecture as art is developed from urban life and serves its surrounding context. Adapting to the environment is the responsibility of design.

Diversity of choice and convenience is fundamental to the nature of cities. The very popular modernist movement from the last century was the cause of a spatial quality that was anonymous and indifferent. Contemporary architects are now attempting to bring back the city's potential and individuality thereby letting the silent city resume its dialogue with its inhabitants.

The city is an organic organism. Every architectural component maintains its own individual voice like those in a choir where some members are mature, some are young and lively, some are elegant and tame. Each voice has its own special quality but at the same time that single voice needs to coordinate with the sound of a group. Perhaps this analogy can describe a harmonized city image.

Rem Koolhaas provided us with another dimension of thinking. Facing the modern city and producing super sized building structures he proposed the concept of "island." Each super structure competes with others for its own territory within the city. The definition of the public domain is questioned as each structure is self contained. Each "island" is the product of individuality. They no longer rely on others to survive; therefore they disengage from the traditional city framework. This is a doomsday science fiction vision of the city but it certainly is a valuable philosophy for an architecture that is accompanied by advanced technology and unrestrained expansion of human desires. The change in building size will eventually cause a change in quality, therefore generating a new form of city. Before that end, we still have an opportunity to choose.

Overcrowding and Loneliness

With limited space and a constant increase in population, consumer products, trash and "overcrowding" now accompany the urban dweller almost every single day. On the other hand, the most common physiological affliction is ironically "loneliness." This is exhibited by lack of emotional connection with others, lack of contact with nature and overall broken spiritual connections.

Therefore the "overcrowding" by apparent material wealth and emotional "loneliness" together kidnap the most fundamental happiness of human beings.

Of course, people have ways to find comfort in things like religion. Once, an anthropologist described India in such a way: In order to survive, every person must keep a vibrant and close relationship with the supernatural. The meaning here is that perhaps to live in the middle of a secular world one must still maintain psychological concentration and be in the right place without losing one's soul or sense of completion.

Can architecture help resolve problems? European churches dating back several hundred years are good examples that prove architecture can actually have some kind of spiritual almost supernatural dialogue with human beings. Modern architects are responsible for "resurfacing" the capability of architecture to move people in ineffable ways.

Order and Chaos

Often buildings published in magazines are considered desirable as long as everything is consistent with the "aesthetic expectations" and situated within an orderly world. Walking towards the "rear" side of any metropolitan city, one has to accept the reality that magazines are not telling the truth.

The terrible chaotic city scenes portrayed in many futuristic movies are actually pretty close to our reality. Civilization's latent self-destructive properties may in fact be reflecting the current violent and chaotic architectural phenomenon.

In this loose, formless, chaotic, fast-paced era only diversity can be accommodated.

This is how Daniel Libeskind describes chaos: "Disorder and randomness originates from a misunderstanding of any type of order and this confusion started from some strange self contradiction that exceeds the range of the order itself." The logic of disorder is a kind of structure, like a mystery with an unreachable truth that foretells the deep attraction of chaos.

Architecture is the conscious reflection of its surrounding environment. In today's declining daily life it is not coincidental that modern architecture is in a state moving towards the unpredictable trend of a "collective chaos", a state that is widespread. Architecture needs to achieve a balance between order and chaos. The ability to adapt to contemporary culture in the presence of a large number of conflicts is the responsibility of today's buildings. Architecture must consider all aspects of this complex world. An architect cannot just pick out what he wants to use to solve specific questions.

Static and Dynamic

The world view of change and impermanence originated in the East and was gradually recognized and accepted by the Western world. From "Metabolist" theory to "Dynamic Form" theory, architects have explored many avenues regarding architecture's ability for self-renewal.

Coop Himmelblau views architecture as an organism: "We hope architecture has more substance. A building is like a person. It bleeds, it's exhausted, its dizziness leads to break downs. Architecture, it glows, it feels pain, cracks under pressure, and weeps. Architecture should be porous (spongy), hot, smooth, angular, cruel, rounded, dainty, colorful, cold, voluptuous, reverent, longing, offensive, moist, dry, even shaking."

Facing our dynamic world, the resolution we provide should also be capable of adapting and continuously growing.

Today's Architecture

Having to retain the memory of history and tradition means also having to maintain an opinion on contemporary culture and allowing the introduction of new "powerful impact" concepts to stimulate a "standing still" common culture. To make progress and to achieve those points is entirely possible. The ruins of history are beautiful and romantic, but they are just creations of the past. Their beauty lies in remembering the glories of the past. This is not a plagiarism of history. Today's architecture should provide beautiful memories of our own era. In most cases these ideas should be "integrated" with many societal aspects in order to achieve a "representation of uncertainty."

Chapter One

全球化与本土化

曾有些恶作剧的人，把美国描述为一个直接从野蛮阶段进入对文化丧失激情的阶段的国家。这也许是文化殖民最极端的一个例子，就某种意义而言是一个文明史上的巨大悲剧。这种文化入侵并非过去时，它依然在地球的各个角落进行着，只是因为两种文明的实力差距没有如此悬殊，给了人类喘息和思考的时间。

在中国，自20世纪80年代开始，西方建筑所产生的强烈冲击随处可见；而与此同时，在全球化和现代化浪潮的冲击之下，那些根植于本土传统的文化诉求，已渐渐达到了前所未有的高度。建筑师有责任展示和传承本土的文化，而不是随波逐流，让植根于土地上的千年文明无声泯灭。

陌生与熟悉

如何在全球化环境中传承本土文化，伦佐·皮亚诺曾提出过一种很有意思的解决方案。他提出"陌生化"和"熟悉化"的概念，认为"陌生化"能激发意识并消解现代文化消费的"催眠效应"。

本土建筑师拥有"熟悉化"的背景这一优势，但长期的单一文化环境可能使其局限于对历史符号简单的"复制"，所以其主要任务是以"陌生化"的批判视角摆脱媚俗。

对于外来建筑师，本身具备"陌生化"的文化背景，所以他们关注的是对地方文化的"熟悉化"学习。

"陌生"与"熟悉"之间，需要实现的是一种"综合再现"。

前卫与媚俗

对于媚俗文化的一味迎合造成了对于优秀文化的漠不关心，这个过程是机械的，就像由公式所控制。媚俗是我们时代中所有谬误的东西的缩影，它耗费了观众宝贵的金钱和时间，却没有满足人类的任何基本需求。它在全世界范围内不分地域和文化界限地抹杀民族文化。

前卫文化，则是为了克服惰性、不停向前而被创造出来的。它必须不断地重新定义自己，它不能保持不变，它的最终目的是为了找到一条在混乱和暴力中使文化不断发展的道路。前卫文化的创造者总是孤立于社会主流之外，但他们却是真正寻求艺术顶峰的勇士。

在当代中国广受地产市场所追捧的新古典主义风格便是对媚俗的最佳诠释，让清醒着的建筑师如坐针毡。

革命与进化

建筑作品需要有"革命"的成分。建筑是通过创造性的探索才超越了构筑物，达到建筑的境界，远比模仿和复制要付出更多创造性劳动。建筑向历史学习，并通过否定而超越历史。建筑因为革命而存在，并以完美为目标。

回顾历史，绝大多数情况下，只有革命性的成就才能在历史的长河中留下深深的印迹，而那些柔和的进化则很容易被遗忘。雅典帕特农神庙、罗马万神殿、巴黎埃菲尔铁塔，都是革命性的成果，都因为具有自己的个性，而成为设计历史上的重要转折点。

对具有创造力的建筑师而言，"革命"是他创作中的重要部分。这里并不是要否定历史文脉的重要性，但是参考过去并不意味着简单复制。当代建筑师应该从过去的建筑作品中汲取教训，并反思如何重现那些存在于过去的复杂性与能量以及如何把它们与当代商业社会、流行文化、时尚、大众传播工具及科学技术中的抽象的隐喻结合起来。

但愿我们对古人的热爱更多些，对他们的抄袭更少些。据说，"希腊人之所以伟大，就是因为他们从不照搬前人所创造的东西。"

符号与语言

查尔斯·摩尔说:"建筑是一种表达艺术。"理查德·迈耶也曾提出人们应该像阅读一本书那样解读建筑物。前几个世纪的建筑师做了很多关于建筑如何自我阐述的深入探索。

矶崎新曾解释他对符号和语义的使用方式:"当我在日本工作时,我用传统的日本感情来表现西方的形式和题材;当我在国外工作时,则又尝试按照西方的逻辑方式来分析日本的建筑形式。"

发展至今,建筑可以是符号,可以是隐喻,甚至可以什么都不是。因而,建筑师更加自由,可以从不同角度来对建筑加以诠释,可以运用各种语言来进行建筑创作。正如卡斯汀·哈里斯所说:"只要你想表达准确的感觉,就无须禁锢我们对语言的运用。"现今我们更关注的是说什么,而不是怎么说。

隐喻与明示

"它们浮现于我们的脑海中,以拾取它们之间可能的联系,就像一个放纵的奢华美梦。"这个关于隐喻的描述证明了一个关于沟通方式的一般规律:即隐喻越多戏剧感越强;而明示越少则越有魅力与神秘感。正如每一个莎士比亚迷知道的那样,具有不确定性的隐喻往往更具表现力。

在建筑学中,解释一个隐含意义反而常常会抹杀了它原有的色彩,这就像分析笑话一样。具有美学意义的实物不仅仅是把它自身提供给我们,而是无言地传达着隐含的讯息。

建筑必须多样化而非只有清晰感,建筑是不确定感与清晰感并存的统一体,建筑就像文章一样属于可读事物,读起来可以激发联想,也能产生回味。

联想仅仅是一方面,通过它一系列有趣的问题被提出来,即使这些问题往往是无意识的、模棱两可的,它们比起理性的、有限制的意识思维更加珍贵,因为它带来"开放式的解读",会赋予这些建筑复杂和广泛的定义。

物质与精神

曾在很长一段时间内,建筑只是被视为人类身体的居所,它作为"精神居所"的功能被全然无视,直到20世纪后半叶才被重新提起。伦佐·皮亚诺曾指出:"只有把技术功能的内涵加以扩展,直至覆盖心理范畴,才能真正使建筑成为人的建筑。"

在处理建筑中的心理学要素时,合理假设某些层面的象征性内容可以被看作是一个出发点。这些象征性内容指的是那些反映环境的历史与文化限定的东西。或许当代艺术先驱杜尚的作品对我们最重要的影响在于它在某一层次上清楚地表达了物体的身份,同时又暗示这种理解并不是唯一的维度。

艺术的重点是将感官刺激转变为对大脑的启发。建筑应具有同样的能力,由物质性引发精神的共鸣。

社会、人、自然

建筑与社会、人和自然三者间的关系在不同的历史时期有不同的表现,时而与自然和谐共处,时而专心于人的使用需求,时而扮演社会的权力象征。回顾建筑史,不难发现,优秀的作品往往是在同一时间成功处理了建筑与三者的关系。

社会在变,人在变,自然也在变,但建筑的任务依然还是于这三者之间找到平衡点。

建筑关乎人类的生活质量,很多新的建筑技术的发明创造是为了提供更舒适健康的微环境和建筑空间,"以人为本"的精神在此得到最彻底的体现。

面对自然的反击,反省和谦恭是建筑可以选择采取的态度,吸取曾经自大妄为的教训,重

Chapter One

新寻找到一种良性的互生关系。

建筑可以被视为一门艺术。只是它涉及更多的政治问题和社会问题，这些是建筑师无法回避的，在这个世界上不存在理想的乌托邦，所有的建筑都必须融入"世俗"。

建筑与城市

城市是一个不断发展的例证，每个城市都有自己的混沌原理。它是一种整体的艺术组合，向世人真实地展示着我们的生活环境。

城市向我们述说了整个社会的复杂的结构——它对权力的态度、对文化和自然的价值的评估。城市证明了一种纯粹的建筑观是没有什么作用的。最好的建筑艺术是从城市生活的文脉中生长起来的，并且为周围环境服务。适应环境是设计的责任。

多样性和多选择性是城市的本性，风靡20世纪的现代主义却造成空间的无名性和匀质性，当代建筑师尝试重新找回城市的生命力和个性，让沉默的城市重新开启与城市居民的互动对话。

城市是一个有机体，组成它的每一个建筑都在发出自己的声音，犹如一个合唱团中，有些人年长稳重，有些人年轻活跃，有些人温柔高雅，因此每个人的声音都有自身的特点，同时又要与其他成员相协调，这也许可以勾勒出一幅和谐城市的图景。

瑞姆·库哈斯给我们提供了另一种思考的角度。他针对现代城市孕育出的超大型建筑单体，提出了"岛"的概念。超大尺度的个体在城市中相互争夺地盘，公共领域的意义受到质疑，每个个体成为自给自足的"岛"。每个"岛"都是个人主义的产物，它们不再依赖对方生存，因而就脱离了传统的城市框架。这是一个末日科幻小说中的城市观，但确实是对伴随技术发展和人类欲望而无限膨胀的建筑体量的一个有益的思考。建筑体积的量变终将会引起质变，产生全新的城市肌理。只不过对于结局，我们依然还有选择的机会。

拥挤与孤独

有限的空间、不断增加的人口和消费品使"拥挤"、垃圾伴随着都市人每一天的生活。而另一方面，这个时代人们最大的心理苦闷是"孤独"，反映在缺乏与他人精神上的交流，缺乏与自然心灵上的接触与交流。物质的"拥挤"与精神的"孤独"一起剥夺了人类最原始的快乐。

当然，人类有自救的方法，比如宗教。有一位人类学家曾这样描述印度：为了生存下去，每个人必须和超自然保持一种非常强烈又非常切身的关系。这里所要说的也许是在芸芸众生中仍能保持的一种精神上的专注，不丢失灵魂的归属感和完整性。

建筑是否也能解决这些问题？几百年前的欧洲教堂就是很好的例子，证明建筑确实能和人类进行某种精神层面的超自然的对话。当代的建筑师有责任"再现"这种建筑对人类的不可思议的抚慰作用。

秩序与混沌

那些刊登在杂志上的建筑图集看上去个个都令人神往，无论从任何事物的角度来考量都符合"期望中的美学"，它处在一个充满"秩序"的世界中。但只要去世界上任何一个大城市的"角落"走一走，便不得不接受"杂志都是骗人"的事实。

很多未来主义影片中所表现得那种"可怕的混沌城市"的场景，确实已与现实十分贴近。文明的这种"自毁"形象的属性可能会成为当前建筑学中狂暴和混沌现象的真实写照。

在这个松散的、混沌的、快节奏的时代，只有多元化才能适应。

丹尼尔·理伯斯基曾这样描述无序："无序和随意源于一种秩序的错乱，而这种错乱出于某种奇怪的超出了秩序本身范围的自相矛盾。"无序的逻辑在于一种结构，像一个神秘的、不可获知的真理那般，已经预示了混沌的深深诱人之处。

建筑是对它周围环境的一种自觉地反映。在日益失序的当今世界，现代建筑形式多种多样、不可预测，由此形成了向"综合混沌"状态转变的趋势，而这种趋势正不断蔓延。建筑必须达到一种秩序和混沌之间的平衡。建筑必须能够适应当代文化中存在的大量冲突。建筑必须考虑复杂世界中的方方面面，而不能只是建筑师挑选出来的他所愿意去解决的特定问题。

静态与动态

变化和无常的世界观源于东方，后来也渐渐为西方所认知和接受。从"新陈代谢"理论到"动态构成"理论，建筑师们对建筑的自我更新能力做了很多探讨和尝试。

蓝天组把建筑视为一个有机体："我们希望建筑物能拥有更多内容。建筑物就像一个生命体，它流血、它枯竭、它眩晕甚至碎裂；建筑，它会喜悦、会疼痛，它在重压之下裂开、流泪；建筑应该是多孔的（海绵状的）、火热的、光滑的、有棱有角的、残忍蛮横的、圆润的、纤细的、丰富多彩的、令人厌恶的、骄奢淫逸的、令人遐想的、引人向往的、使人反感的、湿润的、干燥的甚至颤抖的。"

面对动态的世界，我们所提供的解决方案也应该具有自适应能力，并且能够不断完善。

当代的建筑

秉承对历史和传统的传承，推动当代文化的传播，大胆引进全新的具备"强大冲击力"的概念以激活"停滞"的公众文化，使其有所进步，我们坚信通过不断地努力和进步当代建筑一定能实现时代所赋予它们的上述使命。历史的遗迹是美丽又浪漫的，但是它们只是属于过去的创造。它们的美丽在于对过去辉煌的记忆，它们不是对历史的抄袭。今天的建筑应该能够为我们的时代提供美丽的记忆……这些观念在大多数情况下应该"综合"社会的许多方面以实现"再现的不确定性"。

Chapter Two

CHAPTER TWO 第二章

Tales of Inanimate Objects

The beginning and end of the existence of inorganic objects is almost consistent with the universe. A drop of water or a grain of sand contains the wisdom that inspired many generations of philosophers.

Inanimate objects hold the secrets of the universe in steady silence and are the inspiration for many architects.

无机物的故事

无机物,它们和世界一起开始,和世界一同毁灭,它们几乎是这个地球上可以称之为永恒的东西。尽管没有生命,一滴水,一粒沙中蕴含的智慧却滋养了一代代的哲人。

古老的无机物掌握着宇宙的奥秘,安静无声却亘古不变,在建筑创作中,它们给了建筑师很多启示。

Tales of Inanimate Objects

Chapter Two

Scholar Stone 太湖石

谁怜孤峭质,移在太湖心。出得风波外,任他池馆深。不同花逞艳,多愧竹垂阴。一片至坚操,那忧岁月侵。

——《太湖石》(宋)曾几

Scholar Stone by Zeng Ji, Song Dynasty

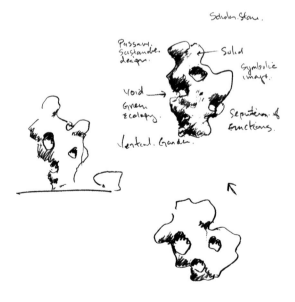

Xi'an Jiaotong-Liverpool University Administration and Information Building

The legends of the Chinese classical garden gave fame to the city of Suzhou. As portrayed in paintings and poems, the artful composition of different landscape elements distinguishes Suzhou's gardens from others by mimicking natural sceneries of rocks, hills, and rivers, providing a poetic utopia and a spiritual shelter for Chinese scholars.

The "Scholar Stone" is an important landscaping element in the Chinese garden. The irregularities of the stone's shapes, colors, textures and sizes have been created by natural sedimentation of the lake, thus giving it an unassuming individuality and whimsical characteristic that differentiates one stone from the other. What makes the Scholar Stone particularly alluring is its potential to manifest ideas and possibilities, as time itself is instrumental in the process of the stone's formation. These abstract figures may be reminiscent of a person, an object, as the primitive simplicity of the stone may evoke a stirring sense of emotion, spirituality, and inspiration. Consequently, many Chinese scholars like to use these rocks as a locus for meditation and contemplation of nature and time.

Situated at a prominent location within the University, this Administration and Information building embodies the idea of Scholar Stone in its materiality and solid-void relationship. Essentially, the building's exterior form is a contextual interpretation of Scholar Stone, while the organic voids programmatically define its building functions.

This building was inspired by a Song Dynasty scholar, Du Wan's poem: "From the essence of heaven and earth, formed as stone out of terrain, without an ordinary shape."

苏州，以中国古典园林闻名于世。正如诗画描述的那样，苏州园林通过仿建石、山、溪的自然景色，打造了不同景观的艺术元素，创造了乌托邦般的诗情画意，长久以来中国文人在此寻求精神的庇护。

太湖石因其多变的体量、颜色、形状以及质感，成为苏州园林中的一项重要景观元素。由于常年在河湖中经水波冲击，石上产生许多孔洞，形状奇特峻峭。独特的外观激发想象引人玩味，"皱、漏、瘦、透"之美与文人精神产生共鸣，其生成和加工过程更是能引发人们对自然对时光流逝的无限遐思。

项目为高校行政信息楼，位置显要，功能复杂。作为一栋坐落苏州的文化建筑，与"太湖石"在精神气质和地域文脉上都有很好的契合。我们以太湖石内部空间的组织形态为概念，建筑内部各功能空间既相互独立又通过虚的空间（孔洞）将各部分有机地组织在一起。内部错综复杂的孔洞暴露在立面上，实现了公众趣味并满足建筑所必须解决的采光、通风、交通问题。

宋杜绾曾云："天地至精之气，结而为石，负土而出，状为奇怪"，也许某种程度上可作为这座建筑的诠释。

Chapter Two

As a result of the city's economic reformation and efforts to globalize education, Xi'an Jiaotong-Liverpool University was established as an international world-class education center within Suzhou Industrial Park. The Administration and Information Building is sited in a key location adjacent to two main promenade axes, with main entrance on the south, the laboratories located on the east side and the classroom building on the north, serving as a dynamic public space for faculty and students.

西交利物浦大学位于苏州工业园区,是一所中英合作创办的综合性大学。行政信息楼紧邻规划中的校园景观轴线,南侧紧邻校园主入口,东侧为实验楼,北侧为教学楼,大楼建成之后将成为一个师生聚集的场所。

Xi'an Jiaotong-Liverpool University Administration and Information Building
西交利物浦大学行政信息楼

Concept Diagrams (Stone - Building)

Location: Suzhou, China 项目位置：中国 苏州
Type: Institution 项目类型：教育设施
GFA: 59,893 ㎡ 项目面积：59,893 ㎡

Awards :

MIPIM Asia Awards 2012, Gold Winner of Best Chinese Future Project

China's Outstanding Architectural Design & Planning Award 2011, Gold Winner of Best Design

Asia Pacific Commercial Property Awards 2009, 4-Star Architecture Award

A' Design Awards 2013-2014,
Winner of Platinum Prize for Educational Buildings - Universities Category

SCMP Chivas 18 Architecture and Design Awards 2013-2014,
Winner of Grand Prize, Winner of Best Public/Community Building in Greater China

Chapter Two

Xi'an Jiaotong-Liverpool University Administration and Information Building

The concept of internal voids of the structure was derived from those of the Scholar Stone's, strategically cut out to correspond accordingly to programmatic requirements. It is no longer a monolithic, static, impenetrable object, but a permeable, breathing mass full of life. By carving out the interiors, the building encourages interaction with the public by inviting people to circulate in and out of the building.

建筑设计理念源于太湖石的孔洞构造，内部的空间因暴露而有了新的意义，并与周围环境有了更加密切与直接的关系与互动。因此，如太湖石般的建筑不再是一个只能被欣赏的孤立的个体，而是以一种更积极的姿态矗立于校园之中，吸引人们徜徉其中。

Chapter Two

Xi'an Jiaotong-Liverpool University Administration and Information Building

The building functions include an administration center, a learning and resource center, a training center and a student activities center. The programs are connected by the circulatory voids inside, consistent with the concept of a Scholar Stone.

行政信息楼由行政中心、学生信息中心、培训中心以及学生活动中心四部分构成，功能复杂。设计上，借鉴太湖石内部空间的组织形态，通过"虚"的空间将建筑内相互独立的不同功能空间有机地组织在一起。

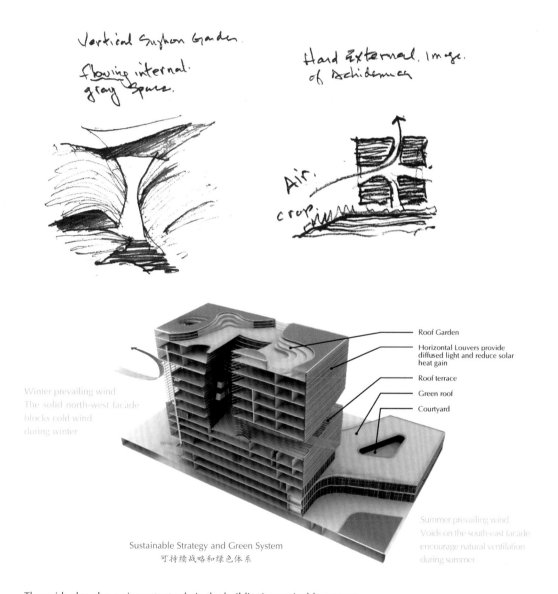

The voids also play an important role in the building's sustainable strategy.

孔洞的概念，为建筑的绿色环保设计提供了有利的支持。

Chapter Two

The unfolding elevation evokes an image similar to that of an ant's nest.

展开立面呈现一种蚁穴般的奇妙构造。

Xi'an Jiaotong-Liverpool University Administration and Information Building

Chapter Two

Xi'an Jiaotong-Liverpool University Administration and Information Building

Chapter Two

Lucky Cloud 祥云

始于地,升至天,天地共融,和谐兴盛。

— *Ancient Chinese Essay about Cloud*

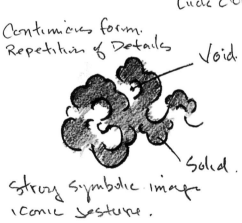

Xuzhou is another historically prominent city which has more than 5,000 years of civilization as well as a reputation as an "Imperial Emperor's Town". We tried to find a link connecting the history of this ancient city with contemporary urban life.

The "Lucky Cloud" pattern was produced in the late Zhou Dynasty and the usage of the pattern was then widely spread throughout the country. Its popularity reached a peak during the Qin and Han Dynasties. The pattern development has some historical understanding with Xuzhou. The cloud is magical and delightful; the changing of the natural form is charming and inspires reverie; the sky above is disguised by clouds inspiring boundless fascination. In this ancient view, the cloud is the symbol of auspiciousness and God's creation. In contemporary interpretations a cloud is a parable for the wild and chaotic dynamics of urban life.

This project borrowed the meaning of "clouds" to achieve connections between time and space. Ancient history, modern and future prospects, and the integration of a built up image within a ringed shape, represented the formation of the atmosphere. The simultaneity of hard with soft symbolized the co-existence of strength and elegance.

徐州，又一座历史古城。华夏九州之一，5,000多年泱泱文明史，更有"一代帝王乡"之誉。我们尝试寻找到一条纽带，把古老的城市历史与当代的都市生活紧密联系起来。

"祥云"纹样产生于周朝中晚期，到秦汉时已遍布全国，达到了鼎盛。"祥云"的发展与徐州有着某种历史的默契。云气神奇美妙，引人遐想，其自然形态的变幻有超凡的魅力，云天相隔，令人寄思无限。在古人看来，云是吉祥和高升的象征，是圣天的造物。置于当代，云又暗合了都会中狂野而充满活力的混沌。

项目借用"祥云"的意向完成了一次跨越时空的结合。古老的历史，现代的步伐以及对将来的展望，融合于环状上升的建筑形象，不仅大气，且刚柔并济，是力量与柔和共存的象征。

Chapter Two

Xuzhou Suning Plaza

徐州苏宁广场

Location: Xuzhou, China
Type: Mixed-use
GFA: 347,321m²

项目位置：中国 徐州
项目类型：综合体
项目面积：347,321m²

Located in the heart of Xuzhou's commercial district, Suning Plaza is a commercial complex consisting of a high-end shopping mall, an office tower, a 5-star hotel, and serviced apartments.

本项目位于徐州市最繁华的中心地段，彭城广场东侧，周边为黄金商业区。项目融合高端购物中心、甲级写字楼、五星级酒店以及高档酒店式公寓，旨在成为立足徐州辐射周边的多功能商业中心。

Chapter Two

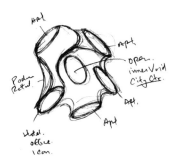

Although the site is set as three individual parts, the master plan synthesizes them into a single entity that links the dispersed commercial podiums. The placement of an alleyway organizes internal circulation along the north-south axis.

The tower on top of the commercial podium reaches 266m; it is a single, iconic entity that contains an office building, serviced apartment blocks, and a 5-star hotel.

项目规划用地分为三块。基地内南北向有一条巷路作为内部道路使用,总平规划将三块地块统一进行考虑,当作一整块用地进行统一规划,商业裙房连通,加强项目建筑的整体感。

商业裙房上的主塔楼达到266米,融甲级写字楼、酒店式公寓、五星级酒店于一体,成为引领整个项目的标志性建筑。

The design is composed by a series of curves to express a fluid and minimalist architectural language. In plan, the podium and the towers manifest a sense of continuity and mobility that enhances the commercial atmosphere. In elevation, emphasis is placed on the smooth transition between the elevating ring-shaped edges and the horizon; its linear formations facilitate the eye into an infinite, seamless trajectory.

建筑造型以平滑的弧形立面为主,追求简洁、流线的建筑风格,裙房与塔楼从设计上来说呈现出流畅、平滑的视觉感,意在创建动感和令人兴奋的商业气氛。立面上以水平线条为主,强调环状上升的趋势,平面流畅的线条使得视线随建筑造型无限延伸。

A sunken commercial plaza maximizes the area of underground commercial and public spaces. The plaza not only addresses the challenges posed by an elongated site, but it revitalizes the commercial strip at the same time by linking the interior with the exterior. Additionally, the sunken center extends land occupancy, enhancing the commercial value of the ground floor.

结合内部道路,规划一个下沉式商业广场,充分结合地下商业和城市公共空间,既解决了地块进深过大的不利因素,又使得本项目内外结合,全面激活了商业气氛。同时,中心下沉式广场也提升了土地利用率和地下一层的商业价值。

Xuzhou Suning Plaza

Chapter Two

Chapter Two

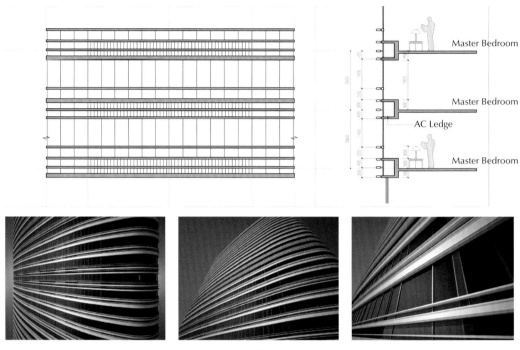

Horizontal Louvers
水平百叶窗

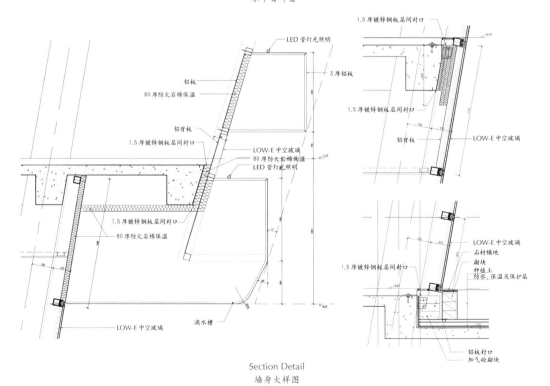

Section Detail
墙身大样图

Xuzhou Suning Plaza

Chapter Two

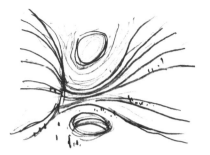

Chapter Two

Landscape　山水

知者达于事理而周流无滞，有似于水，故乐水；仁者安于义理而厚重不迁，有似于山，故乐山。

——《四书章句集注》（宋）朱熹
Notes on The Four Books by Zhu Xi, Song Dynasty

"Landscape" is a recurring motif in this proverb regarding the three evolutionary stages of the man's state of mind:

The first state: Mountain is mountain; water is water (Perception is literal).

The second state: Mountain is not mountain; water is not water (Perception is reinterpreted. Different perceptions are recognized).

The third state: Mountain is still mountain; water is still water (Transcending mere perception in full understanding of an actualized state of mind).

A profound wisdom of life and the universe lies behind what appears to be a game of the words. This ancient Chinese saying is one of many to employ "mountain" and "water" as metaphors in the process of meditation and self reflection, or as objects to be observed in order to gain insights to the mechanisms of nature and the universe. The motif of landscape can also be seen in the Japanese "dry" gardens, in Chinese poetry and abstract paintings. Through the unique understanding of landscape, many Eastern thinkers found their paths to enlightenment.

The project is one dense urban complex of high-rise, striving to engineer a holistic symbiotic environment. The form of the architecture is an interpretation of the distinct natural properties of the "mountain", heavy, grounded and static; as well as those of "water", light, fluid and active. Architectural dialogues are exchanged between the two structures through building materials, form and function; their active interaction and collaboration create a dynamic yet harmonious whole.

先让我们来看古人是如何用"山水"来表达人生三境界的。

第一境界：见山是山，见水是水。
第二境界：见山不是山，见水不是水。
第三境界：见山还是山，见水还是水。

看起来只是文字游戏，背后却藏着深邃睿智的宇宙观。这个例子可以体现长久以来"山"和"水"所具有的独特文化地位，成为东方人在审视自身和自然过程中的重要对象和媒介。日本园林特殊的"枯山水"技巧，中国文人的山水诗画，寄情"山水"，仰天俯地，冥想宇宙，内省自身，成为很多东方思考者的悟道方式。

在这个超高层城市综合体中，我们想要表现一种融合的大气象。"山水"成为很好的载体，两个主要的建筑单体被赋予了"山"和"水"所代表的两种截然不同的特性：一柔一刚；一动一静；师法自然。两个主体建筑间通过建筑材料、形态和功能相互联系，积极互动，形成一个动态和谐的整体。

Chapter Two

333 Shunjiang Road Mixed-use Project
门里成都顺江路 333 号综合体项目

Location: Chengdu, China 项目位置：中国 成都
Type: Mixed-use   项目类型：综合体
GFA: 298,000m²   项目面积：298,000m²

333 Shunjiang Road Mixed-use Project

Sited next to Jinjiang riverbank, 333 Shunjiang Road Mixed-use Project consists of hotel, residential, and apartment towers. Focus is placed on finding a relationship between a large-scaled structure within its natural landscape. The design concept is derived from an image of running stream that cascades through a mountain valley. The residential complex is designed in a fluid configuration to express trickling water. On another hand, the hotel tower facade is arranged vertically to symbolize a mountain's uncompromising grandeur.

门里成都顺江路333号综合体项目位于成都最具城市风情的锦江之畔，集酒店、公寓、住宅及相关配套设施等功能于一体。如何处理体量如此巨大的建筑与城市以及自然的关系是设计面临的首要问题。经过多方案比较，确定了最为合理的结构方案。立面处理则从山与水中汲取灵感。住宅塔楼以柔和的横向线条营造出谷瀑布的水流之感，酒店塔楼以锋利的竖向线条营造山之伟岸大气，两者相得益彰，山水共融。

Chapter Two

The building is sited at a privileged, accessible location in the city, with the north side connecting with a trendy community, the south side opening up to the view of Wangjiang Park, and the west side facing the University of Sichuan campus. Upon completion, the project has the potential to become one of Sichuan's most luxurious commercial and residential communities.

项目北侧与城市时尚生活区相连，南侧临近望江公园，西侧与四川大学隔河相望。项目位置优越，周边社区成熟，配套完善，建成后将成为成都最高端的生活与消费社区。

In response to the limitations of the site, the building is configured to avoid clashing with surrounding buildings. At the same time, the configuration opens up to the view of the river, allowing spaces within to frame the landscape.

由于地块的限制条件极其苛刻，经过大量的分析研究和方案比较，最终的布局既避免了对周边建筑的不利影响，又最大化地利用了周边丰富的景观资源，使得更多内部空间可以一览开阔的滨江与城市景观。

The facade undulates in a subtle way to give the towers a sense of fluidity and gracefulness. The towers are segmented into strips that split and merge at the top and bottom, transforming the mass into a dynamic yet elegant structure.

通过运用现代的曲线造型，为庞大的建筑体量创造了愉悦的视觉效果，利用从上至下分隔的线条使塔楼整体具备变幻的质感，从而创造动态且优雅的形式。

333 Shunjiang Road Mixed-use Project

Chapter Two

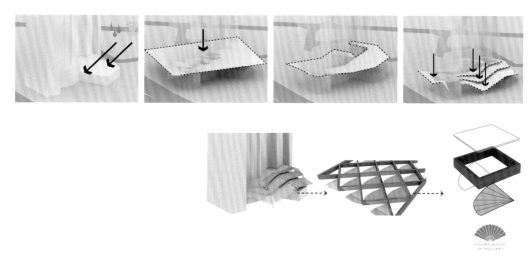

The sales headquarters is formally open and continuous. The roof top opens out through a gradual transformation on vertical louvers. The Celadon appearance is emphasized through a series of twists and distortions on the building's exterior.

售楼处造型简洁流畅，顶部采用未来风格的大幅度曲面立体造型，通过对立面竖向百叶角度的调节，制造出体型扭转的效果，展示出本项目塔楼外立面的风格。

Located on the western side of the hotel, the Art Center bridges the Jinjiang River. A familiar scene to the locals, the ground level is divided into two sloping paths, one for pedestrian and the other for bike circulation. The top of the bridge is covered entirely in ETFE transparent roofing panels, opening the whole level to be read as a public space.

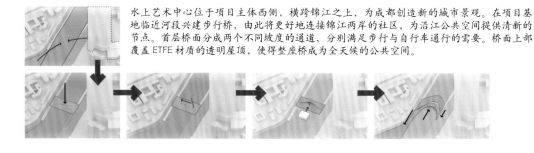

水上艺术中心位于项目主体西侧，横跨锦江之上，为成都创造新的城市景观。在项目基地临近河段兴建步行桥，由此将更好地连接锦江两岸的社区，为沿江公共空间提供清新的节点。首层桥面分成两个不同坡度的通道，分别满足步行与自行车通行的需要。桥面上部覆盖ETFE材质的透明屋顶，使得整座桥成为全天候的公共空间。

To expand on the use of the bridge, a floating oval-shaped banquet hall is placed as an event space for the Chengdu International Research Center, to satisfy the city's administrative demand for political and social events.

桥体中央设置有悬浮的蛋形多功能宴会厅，赋予此设计以更重要的社交功能。此宴会厅将为成都国际研究中心提供会议活动场所，也可以满足城市特定的政治与社会接待活动的需求。

333 Shunjiang Road Mixed-use Project

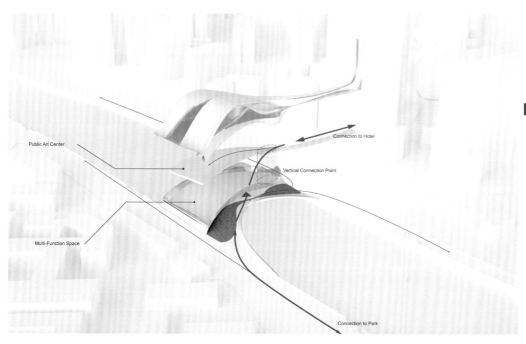

Chapter Two

Chapter Two

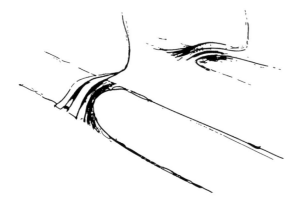

The Art Center derives its form from graceful egrets with spreading wings. Smooth and dynamic, the body of the bridge appears to be rising out of the water in an ethereal, airy motion, creating a world-class architectural landmark for Chengdu.

水上艺术中心的造型灵感来源于锦江上常见的白鹭展翅的情景。桥梁整体一气呵成，在平稳中展现动感。艺术中心的体量似乎要升腾于水面，展现出轻灵飘逸的风格，必将为成都增添一座世界级的建筑地标。

333 Shunjiang Road Mixed-use Project

Chapter Two

333 Shunjiang Road Mixed-use Project

The Mandarin Oriental Hotel Group envisions Chengdu as a contemporary garden city. Its ecological urban landscape does not only aesthetically enhance the riverfront sceneries, it also plays an important role in shaping Chengdu into a thriving, sustainable metropolis.

文华东方酒店集团以全球视野，根植成都现代田园城市的内蕴，营建世界生态田园城市新景观，不仅为锦江设计出最美配饰，也将成为展示当代成都可持续建设的力作。

Chapter Two

River Pebble   鹅卵石

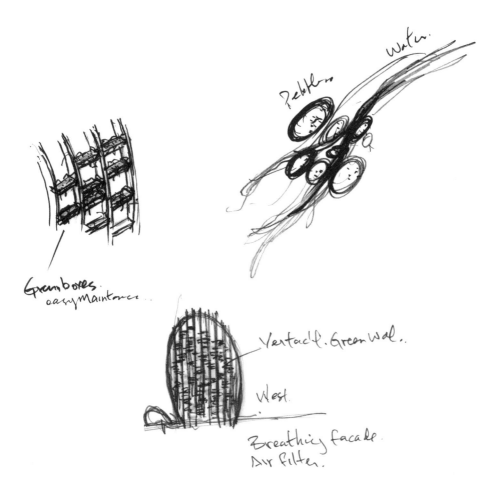

Taipei Nangang Office Tower

In the beginning, all organic forms were rendered in spontaneous acts of nature. Serving as objects of wonder and inspiration, these forms would later be studied, analyzed and reproduced by man. Architecture became a medium for mankind to conceive its own forms in an attempt to generate spatial experiences, to communicate thoughts, to create patterns of behavior.

The river pebbles are one of such forms. Its "egg-like" shapes are eroded by constant streams of floods and rain. Perpetually exposed in time, their forms are, essentially, a narration of unique histories and experiences. Thus, these forms became man's symbol of relentless experimentations and innovative thinking for scientific developments.The pebbles closer to the riverbanks are enveloped in coats of living earth. The growth of moss and other green organisms adds diverse layers of textures to these stones.

The new office is located along the Jilong River, positioned as a natural pebble marked by sets of vertical greens dressing its facade. The new building's "egg-like" shape evokes an initial stage of life; it stands as an incubator of knowledge and a metaphor of intellectual revival, a pebble roaming from the river into the seas, and a symbol of Nangang in its new economic voyage.

自然之手雕琢世间形态，催化万物萌生。作为人类活动的载体，建筑正日益以其丰富多样的空间品质，潜移默化地影响着人们的思维和行为模式。

河滩之石经历溪流雨露的冲刷洗涤，日积月累承受雕琢历练，成为历史的叙述者。接近河岸的卵石往往覆着地衣，轻薄的绿衣包裹在圆润的卵石之上，形成独特质感。

新办公大楼位于基隆河畔，立面垂直绿化的设置使建筑如同矗立岸边的自然之石。新办公大楼更是知识的孵化器，"卵"作为生命初始的形态，隐喻整个地区的复兴。同时，卵石由河入海的生命历程也象征了南港再次踏上新的经济发展的旅程。

Chapter Two

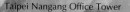

Taipei Nangang Office Tower
台北南港办公大楼

Location: Taipei
Type: Office Building
GFA: 14,169m²

项目位置：台北
项目类型：办公
项目面积：14,169m²

The design for the new Taipei Nangang Office Tower draws inspiration from the shape of the river pebbles, developing a unique aesthetic concept that conveys the idea of roundness and elegance, as well as strength and character.

Located in close proximity to the Jilong River and a major highway in Taipei, the site provides an opportunity for a design proposal that will redefine Taipei's rapidly developing skyline.

台北南港办公大楼的设计灵感来源于鹅卵石，独特的造型传达出圆润和优雅的美学理念，同时又兼具力量和个性。

项目北临基隆河，地处连接新规划科技园区与台北市中心的主要公路旁，位置显要。项目的建设将重新定义这个快速发展区域的天际线。

Chapter Two

Glazing Outdoor Corridors Layered Green Patternized Breathing in

Incubator Knowledge

The 18-story office building has been conceived as an "incubator of knowledge" where innovative ideas are exchanged and turned to reality. The rounded silhouette of the north and south facades tapers to the top, using straight glass panels to rationalize the geometry and to optimize the construction process. The curtain wall wraps around the exterior structure, creating a series of outdoor balconies to allow occupants to enjoy the unparalleled views.

知识孵化器

18层高的办公楼被设想作为一个"知识孵化器",创新思想在这里交流和实现。南北立面的弧形轮廓线随着高度增加而收窄,采用合理的几何优化策略,尽量使用平板玻璃幕墙实现立面效果,简化施工程序,降低总体造价。幕墙系统与外部结构相契合,创造了多个室外阳台,为使用者提供无与伦比的观景条件。

Urban Living Room

The plan of the office space aims to provide an efficient, healthy, and collaborative working environment. Communal areas such as kitchens, coffee corners, small libraries, and meeting rooms are proposed as the "urban living room", which aims to promote creativity and to encourage interaction between people.

城市客厅

办公室空间由多个贯穿各层的"城市客厅"串联而成。此处内部空间可依据需求安排咖啡厅、烹饪间、小型阅读区、讨论室等。空间功能的多样性给办公人群提供了多种选择,并有效地优化了办公品质,进而促进思维的融汇交流。

Taipei Nangang Office Tower

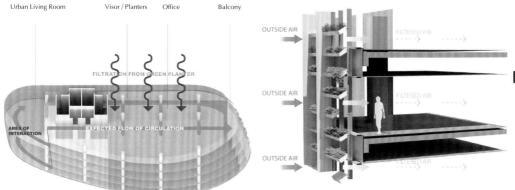

Urban Living Room Diagram
城市客厅空间流动分析

Green Planter System as Facade Filter
绿植墙系统分析

Green Filtration Wall

With the intention of implementing sustainable strategies and obtaining a LEED certification, a breathable building envelope was applied to the building to add diversity to the facade, as well as to moderate heat gain throughout the year. The vertical aluminum fins have a specified distance and depth to control the amount of sunlight penetrating into the building. Green planters are placed on the west side to provide sufficient shading and to regulate the interior temperature during summers.

植栽式表皮
建筑外皮通过应用可呼吸式的表面形式,增强了外观多样化,同时调节了室内热量的获取。立面纵向鳍状飘板及西侧绿植协同合作,有效控制建筑热耗,从而实现低碳节能的设计目标。为实现可持续的发展规划,此项目已将申报美国LEED绿色建筑认证提上日程。

Chapter Two

Taipei Nangang Office Tower

Original Massing
初始体量

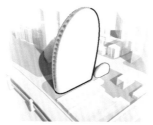

North: Leisure Balconies
北面：休闲活动露台

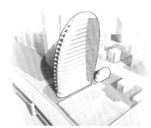

South: Urban Living Room
南面：城市客厅

Chapter Two

Jade 玉石

折琼枝以为羞兮，精琼靡以为粻。为余驾飞龙兮，杂瑶象以为车。

——《离骚》（战国）屈原
Lisao by Qu Yuan, Warring States Period

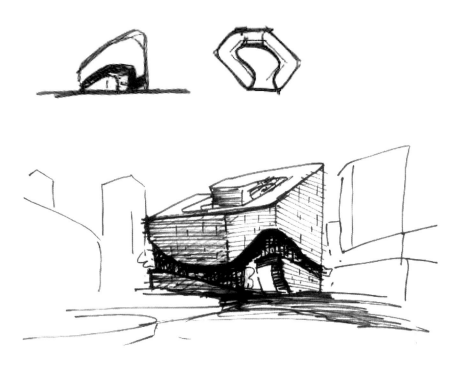

Shanghai EXPO Urban Best Practice Area E06-04 Block

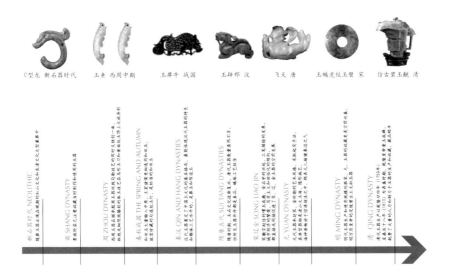

Jade is a stone deeply ingrained in Chinese art, culture and history. It has a long and slow process of geological formation. It has established a presence, as a symbol of wealth and power, that is as ancient as Chinese civilization.

The site has undergone rapid transformation after the convening of the Expo. Now, the area faces the challenge of reintegrating itself back into the urban fabric while engaging the everyday life of its inhabitants. This is the design problem which we address with this project.

The design concept stems from the evolution of Chinese jade culture. The cumulative history of its morphology, both physically and symbolically, generated this architectural form: the embodiment of 5,000 years of Chinese civilization and an open gesture to the future of urban life.

玉在我国的历史可谓源远流长。 在长期而缓慢的进化过程中，玉由原先的仅仅是一种特别性质的石头转化为权力、地位、财富的象征。

项目所在的区域因世博会的召开而经历了一次急速转变，如今这片区域面临着如何重新融入市民生活的挑战，在设计的过程中，必须对这一问题给予回应。

设计概念源于历史悠久的中国玉文化，从玉的历史形态演化过程中汲取灵感，形成建筑形态的生成理念，在后世博会的背景下既体现中华五千年文明的沉淀，同时又对现代都市生活呈现一种开放的姿态。

Chapter Two

The project is situated at a prime location: at the northwest region of the site for Urban Best Practices Area of Shanghai World's Expo, adjacent to the city's arterial roads. It will become a gateway to the Expo. The project mostly consists of offices, with retail on the ground and second floor. The courtyard opens to the city, serving as an important public space for the community.

此项目位于上海世博会城市最佳实践区的西北角,毗邻城市干道,建成后将成为后世博的门户。项目功能以办公为主,一二层为零售店面,建筑围合的内院对外开放,服务于社区,成为市民活动的重要场所。

Shanghai EXPO Urban Best Practice Area E06-04 Block
上海世博会城市最佳实践区 E06—04 地块

Location : Shanghai, China
Type: Mixed-use
GFA: 50,000m²

项目位置:中国 上海
项目类型:综合体
项目面积:50,000m²

Chapter Two

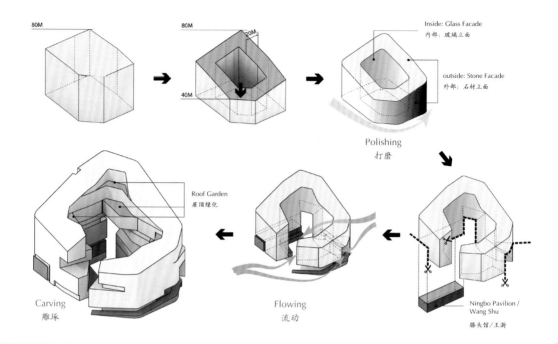

Shanghai EXPO Urban Best Practice Area E06-04 Block

The volume of the building is a reference to the manufacturing process of jade in the techniques of grinding, carving and puncturing, to preserve Wang Shu's Ningbo Pavilion while creating a sense of spatial flow.

建筑体量的衍生参考了玉器的琢磨过程,通过对形体的打磨、雕琢、开洞的手法,保留地块内由王澍设计的滕头馆,同时营造空间上的流动感。

Ecological and sustainable design is fully considered in architectural design and the interior space. Except for exterior greening platform, thin depth tower enhances natural ventilation and utilizes natural daylight. In addition, other ecological measures are also adopted in the design and it will meet 3 star of China Green Building and LEED platinum certification.

建筑充分考虑了生态及可持续发展设计,除室外绿化平台外,薄板楼的设计使自然通风和采光极具优势。此外设计综合采用了其他生态手段,将符合中国绿色建筑三星及LEED铂金级别标准。

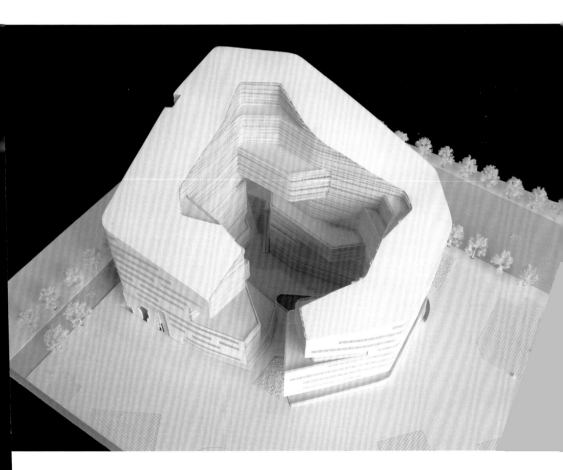

Chapter Two

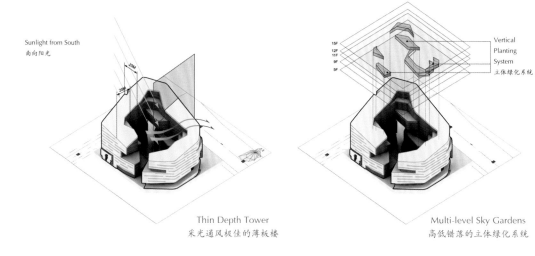

Sunlight from South
南向阳光

Thin Depth Tower
采光通风极佳的薄板楼

Vertical Planting System
立体绿化系统

Multi-level Sky Gardens
高低错落的立体绿化系统

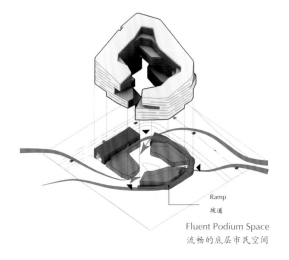

Ramp
坡道

Fluent Podium Space
流畅的底层市民空间

Shanghai EXPO Urban Best Practice Area E06-04 Block

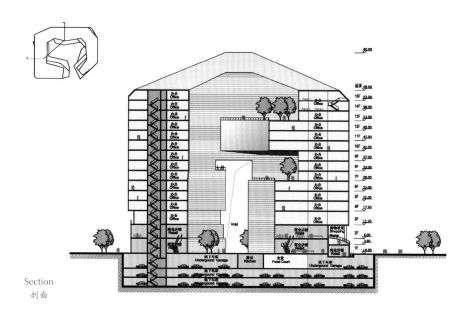

Section
剖面

Chapter Two

River　河

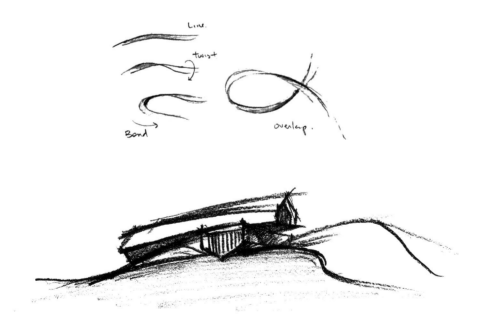

Tianjin Century Times Art Museum

Tianjin—The Celebration of Nine Rivers' Reunion

The City of Tianjin, with 1,500 years of history, was initiated and promoted by the renowned man-made water way transportation project: the Grand Beijing-Hangzhou Canal. It officially became a major city at the time of Yongle, Ming Dynasty. This city left a mark in Chinese modern history, with a mixture of various colonial style buildings along with traditional vernacular architecture establishing the city's identity.

The Wuqing Bashang Art Exhibition Center is located in Wuqing County, City of Tianjin. The site is located right beside the ancient, scenic man-made river: The Grand Beijing-Hangzhou Canal. The Grand Canal, claimed as the longest canal in the world, along with the Chinese Great Wall was considered as the two most marvelous engineering achievements by the ancient Chinese people. From north to south, this canal intersects with the Hai River, the Yellow River, the Huai River, the Yangtze River and the Qiantang River. This man-made water way largely facilitated transportation of goods from the rich Jiangnan region, also enhanced the cultural communication between north and south China, and itself has evolved into an important symbol that represents the Chinese cultural connection.

九河下梢天津卫

天津，初建于一千五百年前京杭大运河开通，因漕运而起，亦因漕运而兴。明永乐年间正式筑城，近代百年天津写下了中国近代史上重重的一笔。而不同殖民风格连同当地的传统建筑一起从那时起就已然确定了这个城市的风貌。

武清坝上艺术展示中心位于天津市武清区，场地坐落于历史悠久、风景怡人的京杭大运河河畔。京杭大运河为世界上最长的人工运河，与长城并称中国古代两大工程。运河自北至南贯通海河、黄河、淮河、长江、钱塘江，漕运江南丰饶物资，融汇中国南北文化，成为中华民族文化纽带的代表。

Chapter Two

Tianjin Century Times Art Museum
天津坝上春秋美术馆

Location: Tianjin, China
Type: Exhibition
GFA: 3,600m²

项目位置：中国 天津
项目类型：展示
项目面积：3,600m²

The design positioned the project as a Cultural and Exhibition Center. Recognizing the Grand Canal cultural significance and the history of the City of Tianjin, the design drew inspiration from the Chinese traditional calligraphy, using the cursive character " 河 " as the central concept.

方案以文化建筑展示中心为定位，深谙运河文化与天津文化，以之出发；并从中国传统书法中获取灵感，把草书的"河"字作为设计起点。

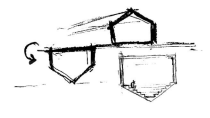

Chapter Two

Architecturally, the building facade uses flowing curves and vivid lines as a metaphor to water; the whole building form implies Chinese calligraphy with elegance.

建筑造型上运用流动的曲线，线条生动亦暗合曲水流觞之意境；而建筑整体又源自中国书法，隐含了浓浓中国文化的翰墨风雅。

To correspond with the inherited merit of the site, at the two ends of the building, the architectural language returns to the very original symbol of a house. With large window openings, the building is able to bring the river and water pond views into the interior spaces.

借助地形优势,建筑形式的末端引用了最纯粹传统的建筑符号,身处建筑的内部也可以饱览场地周边的河景和水景。

Chapter Two

Tianjin Century Times Art Museum

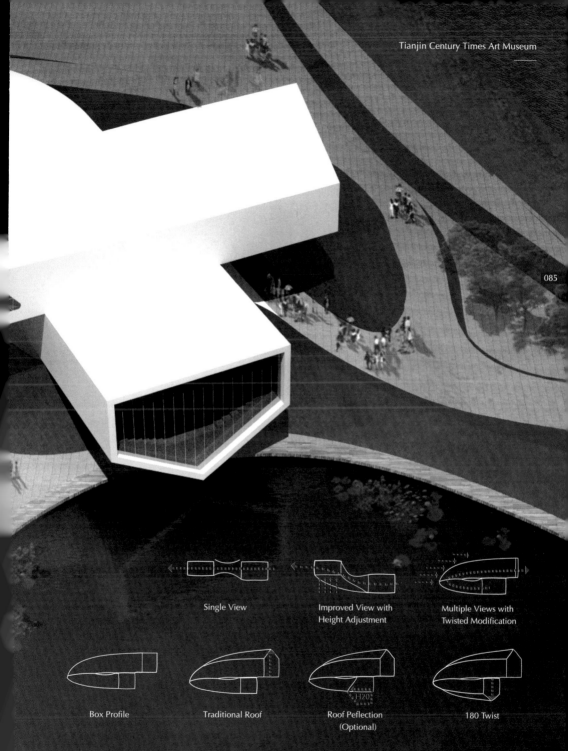

| Single View | Improved View with Height Adjustment | Multiple Views with Twisted Modification |

| Box Profile | Traditional Roof | Roof Peflection (Optional) | 180 Twist |

The circular architectural form of the Tianjin Century Times Art Museum connotes cycle of life which responds to the long-standing Oriental world ideology. It is a dialogue between contemporary architectural language and ancient wisdoms.

坝上春秋美术馆的环状造型体现了循环往复、生生不息的意象，与源远流长的东方世界观相应和，是现代的建筑语汇与古老的人生智慧的对话。

Chapter Two

Alluvial Plain 冲积平原

清浅白石滩，绿蒲向堪把。家住水东西，浣纱明月下。

——《白石滩》（唐）王维

Bai Shi Tan by Wang Wei, Tang Dynasty

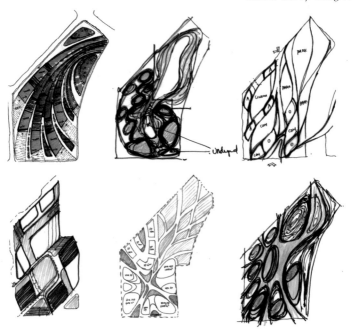

The alluvial plain is a largely flat landform created by the deposition of sediment over a long period of time by one or more rivers coming from highland regions. These lands are fertile, convenient in transportation, thus are well developed economically.

The design draws inspiration from the intricate landscape: a drawing formed by the intertwining rivers and saturated with wildlife, natural and vivid. We seek to apply this relationship and interaction between natural elements as a concept for a large, mixed-use development.

Architecture becomes "plains": solid, grounded and physical; green space becomes "river": airy, light and dynamic. The design introduces a gesture of organic reaction, as the park to the north continues into the project site with the fluidity of the river.

冲积平原，是由河流沉积作用形成的平原地貌，一般位于河流的下游河口，土地肥沃，交通便利，经济发达。

打动我们的是这种地貌所呈现出的美妙的图底关系，交织的河网勾勒出一个个裸露的"平原"，自然生动，颇堪玩味。在现代城市中是否也有可能在综合体的尺度上来模拟这种构图所带来的空间趣味？

尝试将北侧的公园景观如河流一般引入南侧的建设用地，建筑成为坚实、平坦的"平原"，绿色成为穿插其间的轻盈、动感的"河"，打破一般综合体的沉闷体量，使建筑与环境有机地结合在一起。

Chapter Two

Cheng Ying Center

诚盈中心

Location: Beijing, China
Type: Mixed-use
GFA: 92,650m²

项目位置：中国 北京
项目类型：综合体
项目面积：92,650m²

Cheng Ying Center

The project is located in Beijing Wangjing area, near the Beijing-Chengde Expressway, the Airport Express as well as major roads. The 1-5 km radius is rich in landscape: three existing large parks, as well as one under planning on the northeast side. Meanwhile, the site has a number of artistic and creative venues, attracting a variety of consumer groups.

项目位于北京大望京区域，临近京承高速、机场高速等城市交通干道，交通优势明显。周边1-5公里范围内拥有三个大型公园，紧邻基地东北侧为规划中的城市公园，景观资源丰富。同时，区域内有多个艺术创意园区，能吸引多样的消费人群。

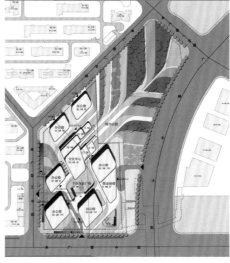

Chapter Two

生成过程
Concept Evolvement

1 80 Meter Zone
80米限高区

2 Overlooking
正向对视严重

5 Lowrise Buildings
低层建筑

6 B1 Commercial Flow
地下商业流线

The design addresses the challenge of how to utilize tight spaces to create comfortable, free flowing spaces. The concept is a break from traditional layouts. The six buildings are rotated 45 degrees, thus capable of receiving the full benefits of landscape resources while retaining its degree of privacy, all the while displaying the most ideal gesture towards the southeast corner of the city square.

如何在相对紧张的用地内营造舒适流畅空间是本案面临的挑战。设计打破传统的建筑布局，将六栋办公楼旋转45度，以获得充分利用景观资源、避免对视，同时也以最优美的姿态面向东南角的城市广场。

Since the project is not located in the urban center with rich natural resources in its surroundings, the offices are designed with more visual and physical connections to the surrounding green spaces. The underground commercial spaces are integrated to the subway station and sunken plaza. The cultural and F&B facilities are placed to the north-east corner of the site , which directly faces an urban park.

由于该项目不处在城市中心区，周边自然资源丰富，因此为办公区设计了大量可供观赏的通道面向绿化区以创造更具有人文与艺术气质的办公环境。通过地下商业空间和地铁站点及下沉式广场相连接。充分利用基地东北侧的城市公园，使休闲、餐饮、办公和文化设施有机结合在一起。

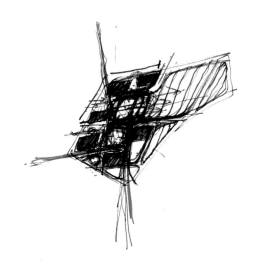

Cheng Ying Center

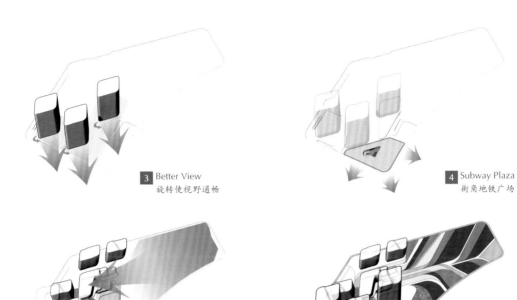

3 Better View
旋转使视野通畅

4 Subway Plaza
街角地铁广场

Cheng Ying Centre 7 Green Connection
引入公园绿化带

8 Masterplan
规划方案

Chapter Two

Chapter Three

CHAPTER THREE 第三章

Tales of Creatures

There are moments in life when we are deeply touched by the sight of nature. Although we live in a world dominated by machines, plants remain as the primary source of life. From its limited lifespan and generations of constant evolution, we learned the change in seasons and the eternal circle of life.

Animals, referred to as friends of the human race, share the planet with us. We are inspired by their beautiful physical form and are amazed by their collective behavior. The study of animals serves as a way of exploring the outside world and even as a way to lead us to a better understanding of ourselves. Animal totems can be found in many ancient arts. Some are still regarded as the symbols of certain ethnic groups and cultures.

生物的故事

路边的一株幼草，风中一片沾衣的落花，总会在某些时刻触动内心最柔软的部分。植物无声地给予人类回归自我的力量，在机械化了的世界，是这些生命提醒我们四季的更替和生命的规律。

人们常常形容与我们共处一个星球的动物为人类的朋友。动物世界的绚烂多姿、动物间的生存竞争，都深深影响着人对世界和自身的认识。从人类的原始信仰中，可以找到很多美妙的动物图腾，或真实或虚幻，有些至今仍是某种民族或文化的象征。

Tales of Creatures

Chapter Three

Hibiscus　　芙蓉花开

疑是春风错采霞，秋江三醉木莲花。红颜偏作金风客，粉面独开后蜀家。莫道荷生泥不染，只言菊隐玉无瑕。锦城秀丽芙蓉放，敢斗寒霜不自夸。

——《木芙蓉》

Ancient Chinese Poem about Hibiscus

Chengdu Jinniu District North 1st Ring Rd. N Project

Flowers are fundamental to man's innate understanding of beauty and life. The blossoming of flowers suggests a metaphorical co-existence of elegance and persistence, and has been a common subject in folklore and poetry. Within painting, sculpture and architecture, boundless reinterpretations of the flower generate a spiritual resonance that moves and inspires.

The hibiscus is the city flower of Chengdu. The origin of Chengdu and the hibiscus flower dates back to the Five Dynasties 1,050 years ago: An emperor favored the hibiscus flowers; he ordered the people to plant the hibiscus along the city walls during blooming seasons and created the famed "forty li of splendid view." Since then, Chengdu has earned the name of "Hibiscus City." Residents of this city seem to have inherited the sentiments of the emperor; despite the changes of dynasties over time, Chengdu still retains its laid-back, leisurely city rhythm.

In this residential project, we hoped to break the traditional model of residential development and explore a more innovative housing form. Considering the climatic conditions in Chengdu, we provided users with a large number of balconies, rooftop gardens, and other public recreation spaces. The building image is an abstract expression of the hibiscus, as we strive to reflect, both internally and externally, the spirit of the city.

花是美的化身，生命的象征。花传达了一种绽放的美丽和坚韧，成为传说和诗歌吟咏的对象。在绘画、雕塑和建筑装饰中，花长久以来都被视为灵感的源泉，被艺术家们一次次地重现，带给人们心灵的共鸣和感动。

芙蓉花是成都的市花。成都与芙蓉花的渊源始于五代十国时一位皇帝对芙蓉花的偏爱。皇帝命百姓在城墙上种植芙蓉树，花开时节，"四十里为锦绣"，成都便有了"芙蓉城"之称。这座城的居民似乎继承了这位皇帝赏玩芙蓉的闲情逸致，任时光变迁，成都依然保留了其悠闲宜居的城市节奏。

在这个以住宅功能为主的项目中，我们希望打破传统的住宅开发模式，探索一种更新颖的住宅形态。结合成都的气候条件，给使用者提供大量露台、空中花园等公共休闲空间，同时其建筑形象是对芙蓉花的抽象表达，力求项目的内在和外在均可与城市精神产生共鸣。

Chapter Three

Chengdu Jinniu District
North 1st Ring Rd. Project
成都金牛区北一环项目

Chengdu Jinniu District North 1st Ring Rd. N Project

Location: Chengdu, China
Type: Mixed-use
GFA: 456,848m²

项目位置：中国 成都
项目类型：综合体
项目面积：456,848m²

Located in Chengdu North First Ring Road, the project is a mixed-use residential and office complex with a goal to enhance the typical housing experience. As a result of Chengdu's relaxing atmosphere and soothing weather conditions, its inhabitants prefer outdoor recreational spaces with pleasant views over traditional building orientations. Considering these factors, a radial plan was devised, implementing central green spaces and maximizing views.

Inspired by the shape of a hibiscus petal, the concept is translated into the architecture by manipulating the edges of balconies to create an abstraction of metamorphosis.

项目位于成都北一环，是一个以高端办公与住宅功能为主的城市综合体。项目伊始，与甲方沟通达成共识，要做一个有别于普通住宅的住宅项目。充分考虑成都的气候和文化生活习惯，成都人喜好户外休憩空间，对景观的追求重于朝向，可以接受住宅东西向。鉴于这些特点，尝试辐射状的布局方式，既提供了中央绿地，又避免住宅间的对视。

建筑造型灵感来源于芙蓉花瓣，运用阳台的变化柔和建筑边界，令人印象深刻。

Chapter Three

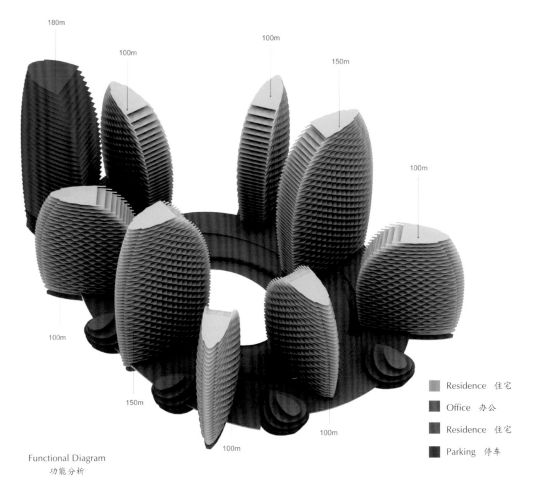

Functional Diagram
功能分析

The residential compartment is connected through a series of circular podiums on the ground floor, opening up a central space allocated to the club house.

One side of the tower is sliced into layers to create a series of vertical green spaces. As for the podium, it features a roof garden and a central sunken plaza. A unified, picturesque landscape is composed by blending these three elements together.

住宅与办公塔楼由环形的商业裙房串联,形成一个整体,在其中央布置会所。

塔楼侧面的层层退进提供了垂直绿化空间,裙房顶部为屋顶花园,环形中央为下沉广场,三者构成立体化的绿化景观体系。

Green Diagram
绿化分析

Chengdu Jinniu District North 1st Ring Rd. N Project

Chapter Three

Chengdu Jinniu District North 1st Ring Rd. N Project

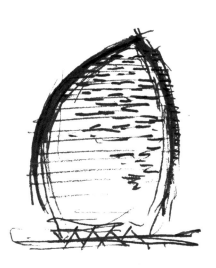

The overlapping, curved geometry naturally form pleasant and practical double-height balconies.

交错的曲线阳台自然而然地形成了两层通高的阳台，美观而实用。

Chapter Three

Oasis 绿洲

交河虽远地，风俗异中华。绿树参云秀，鸟桑戴嫣花。

——《西州》（唐）
Tang Dynasty Chinese Ode to the Oasis

Xiamen Green Land Mixed-use Project

As a geographical term, "oasis" is defined as an isolated area of vegetation in a desert, typically containing a water source. However small the area maybe, the oases provide essential habitats for animals where isolated ecosystems may thrive. Thus, they play a significant role in shaping the surrounding ecological landscape and the microclimate.

This project is such an analogy. In the face of rapid urbanization, natural spaces are constantly being violated, eroded with a certain form of collective urban sickness, in a process similar to desertification.

Humanity fights side by side with nature. The man-made oasis stands as a candle in an ocean of darkness, a seed against a backdrop of barrenness, a sign of life amongst the machines.

How do we face the city? Will nature be the helping hand? Unprejudiced against all, nature governs life in its own spontaneous display of order and chaos, with methods that man cannot fully comprehend. However, as long as we are hard at planting, the oasis will flourish and a resting ground for tired travelers.

In China, the effects of mass urbanization sometimes bare the aftermath of large "disasters". Every city is busy at creating its own iconic, central complex: the ultimate product of urbanization. Due to the diversity and scale of its functions, such a complex can be re-interpreted as an "urban oasis". This time, we plant this concept in the city of Xiamen.

"绿洲"作为地理学名词的定义是：大尺度荒漠背景基质上，以小尺度范围，但具有相当规模的生物群落为基础，构成能够相对稳定维持的、具有明显小气候效应的异质生态景观。

我们尝试做一个类比，城市的快速扩张，与城市化形影相随的各种城市病的蔓延，自然空间被城市侵蚀驱赶，渐渐消失殆尽。某种角度而言，城市化与沙漠化的表现如此相似。

人类在荒漠面前，与自然并肩作战，一个个的人造绿洲，也许渺小，但却是荒芜中的种子，黑暗中的光，一点点的绿色，被寄予的是对生命的期冀。

那我们该如何面对城市呢？是否自然也会助我们一臂之力？公平而无私，这是自然的品性，它有自己的混沌和秩序，在人类看来也许有些无常，却蕴含绵长的生命力量。只要我们辛勤地播种，在城市的荒漠中，"绿洲"也会生长，长成旅人休憩的乐园。

中国，城市化的"重灾区"。每个城市都在打造自己的"城市综合体"，即城市化的终极产物，可同时它又因其功能的多样性，巨大的规模，非常适合成为"城市绿洲"的载体。这次，我们试着在厦门撒下一颗种子，营造一片绿意。

Chapter Three

Xiamen Green Land Mixed-use Project
厦门绿地综合体项目

Location: Xiamen, China
Type: Mixed-use
GFA: 295,000m²

项目位置：中国 厦门
项目类型：综合体
项目面积：295,000m²

The project is located in the old city center of Xiamen, surrounded by a series of highly dense buildings. We seek new ways of inserting energetic growth factor in the form of architecture within this ancient urban context. The programs consist mainly of commercial, retail and hotel.

Here, "illuminating the green city lights" is not just a slogan, but also a driving force for design. This project can be viewed as a green oasis amongst the concrete jungle, introducing nature as an essential part of everyday life, to improve the quality of living for its inhabitants.

项目位于厦门的旧城中心，周边建筑密度高，项目功能定位以商业、百货、酒店为主。如何在一个较成熟的城市文脉中插入一个新的能量因素成为设计的主要动机。

在此"点亮城市绿色之光"不仅仅是一句口号，而是作为整个项目的驱动力，在钢筋水泥的城市环境中创造一片绿意，借助自然的力量来改善人群既有的生活模式，改善居住者生活品质。

Xiamen Green Land Mixed-use Project

Chapter Three

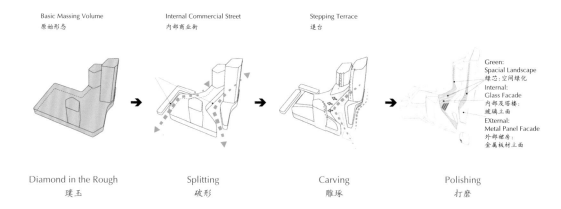

Basic Massing Volume	Internal Commercial Street	Stepping Terrace	
原始形态	内部商业街	退台	Green: Spacial Landscape 绿芯：空间绿化 Internal: Glass Facade 内部及塔楼：玻璃立面 EXternal: Metal Panel Facade 外部裙房：金属板材立面
Diamond in the Rough 璞玉	Splitting 破形	Carving 雕琢	Polishing 打磨

Xiamen Green Land Mixed-use Project

Forming
成型

We aim to create a three-dimensional framework best fit for the existing ecosystem. Interwoven green roofs, green belts and green courtyards become the breathing flesh and blood of the complex, while pathways become skeletons. Together, these elements provide a unique, integrated experience for all users of the park.

项目旨在城市中心打造立体生态系统。以屋顶绿化、退台绿化、中庭绿化为血肉，以屋顶步道、退台通道为骨架，交织组成体验性的空中城市公园。充分利用屋顶绿化、生态空中大堂等提升局部生态环境。

Chapter Three

Xiamen Green Land Mixed-use Project

Chapter Three

Xiamen Green Land Mixed-use Project

Inclusiveness — Metropolitan street and flagship stores
They bring about the perception of a diverse and inclusive Xiamen, combining Eastern and Western cultures with high quality retail products.

— Urban Stage
The public plaza, the people's platform, link culture with commerce.

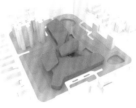

— Street of Lights
On this old retail street, time freezes and tradition is preserved amongst the festivities.

Growth — Hotel & Apartments
Xiamen's growth is also implemented in its rising standard of living and material goods. The site is flourishes vitally at nighttime.

Gregarious — Engaging Financial Exchanges
Future business interactions will occur amongst these "clouds" symbolically; a representation of Xiamen's economic progress.

包容度 —— 大都会名品街和旗舰店
感悟一个多元包容的厦门，糅合东西方都市文明的商业气息。

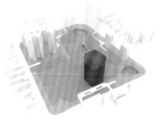

—— 都市秀场
市民的节庆广场，大众的文化舞台，以文化贯通商业文明。

—— 灯火街
老字号商业街，汇集最厦门的老字号，老味道，时光在此流连。

生长态 —— 酒店及酒店式公寓
贯彻一个品质生长的厦门，实现项目整体的夜间生命力。

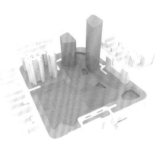

交互云 —— 泛金融产业交互云平台
打造一个云端交互的厦门，承载企业价值梦想。

■ Metropolitan Street 大都会名品街
■ Urban Stage 都市秀场
■ Street of Lights 灯火街
■ Serviced Apartment 酒店式公寓
■ Hotel 酒店
■ Office 办公楼

Chapter Three

Xiamen Green Land Mixed-use Project

Three towers, six faces, all shine brilliantly day and night. Four edges, responsive and inclusive of all components of the urban environment, activate the wayward journey of Xiamen's future development.

三栋高层，六个切面，日也辉煌，夜也灿烂。四方中心，包容城市万象，开启厦门动感魅力的新旅程。

Chapter Three

Tea Plantation　茶园

茶。

香叶，嫩芽。

慕诗客，爱僧家。

碾雕白玉，罗织红纱。

铫煎黄蕊色，碗转曲尘花。

夜后邀陪明月，晨前命对朝霞。

洗尽古今人不倦，　将知醉后岂堪夸。

——《茶》（唐）元稹

Chinese Poem about Tea by Yuan Zhen, Tang Dynasty

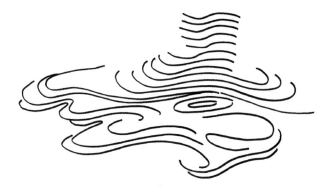

Suzhou Hong Leong City Center

This is an interpretation of "tea" as a custom.

Since the ancient times, "tea" has been linked to the notion of "birth." In this modern information society composed of cold machines and rapid transfers of data, "tea" has become one of those invisible means of communication, but delivered in much direct and intimate manners. "Tea" brings out a softer, more sincere temperament, with a familiar openness which expresses the undefinable space between close friends.

Retail design has a tendency to be vibrant and loud on the exterior, yet impersonal and apathetic on the interior. Tea plantations appear to be organic and gentle, yet possess the magnetic attraction that allows the passerby to take rest. This specific "tea plantation" is not only a space for oneself, but also a space for interaction and exchange between individuals.

By inheriting the typology of a Chinese tea garden, the project introduces in a rich spatial experience both interior and exterior. It aims to interact with the occupants' behavior pattern by providing varied spatial quality.

我们尝试做一次对"茶"的重新解读。

"茶"自古以来都与"出世"的态度相联系，而在如今这个信息可以通过各种看不见的波来传递的时代，喝"茶"反而成了一种更直接的、可视的沟通方式。"茶"那种出尘冷清的气质渐渐转换为一种柔和开放的姿态，如同密友间的开诚布公。

商业项目总令人想到外表的热闹喧哗和内里的空虚冷漠，不能做出一些改变吗？茶园给了我们一个提示，它看起来自然柔和，却透出一种内敛绵长的吸引力，人们可以在此得到片刻的休憩。这里的"茶园"不是一个自省的场所，而是一个可以相互交流的空间。

本案通过模拟茶园的形态，创造出尽可能丰富的室内外活动场所，尝试利用空间形态来引导使用者的生活方式。

Chapter Three

Suzhou Hong Leong City Center

Co-designer: Keith Griffiths

苏州丰隆城市中心

Location : Suzhou, China
Type: Mixed-use
GFA: 295,402m²

项目位置：中国 苏州
项目类型：综合体
项目面积：295,402m²

The site is located in the heart of Suzhou Industrial Park. It is one of the major phases for the development of the city, with surroundings which offer fantastic views. The north side of the site is in close proximity to the green space under planning, while the south side faces Jinji Lake.

The project consists of four functions: apartments, office, hotels in the form of super high-rises, and a commercial podium.

Suzhou Hong Leong City Center

项目位于苏州工业园区的核心地段，是该区域几块重点城市综合开发建设地块之一。基地北侧与规划城市绿地相邻，南侧面向金鸡湖，景观条件非常优越。

本项目集公寓、办公、酒店和商业于一体。由四栋超高层塔楼以及大型商业裙房组成。

Chapter Three

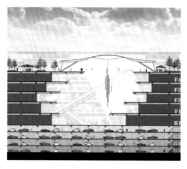

The commercial podium introduces several featured atriums. Shops are cascaded to invite natural light in, providing speedy access to the rooftop.

商业裙房拥有多个主题中庭,店铺叠层跌落布置,有效引导自然光,其中的通天梯将行人带至屋顶花园。

Being a mixed development, it will become a 24/7 vibrant urban center. The project provides apartment, hotel, office, shopping mall, recreational facilities, and outdoor gathering places.

此项目将是一个全天候运作、充满生机的城市综合体,由公寓、酒店、办公、购物中心、休闲设施和户外活动场所等功能组成。

Suzhou Hong Leong City Center

At the podium, a component of the outdoor corridor is introduced on top of a traditionally large scaled retail mall. The retail street on top of the roof contributes to the continuity of commercial circulation and maximizes the efficiency of commercial space.The new typology of the podium enhances its overall commercial value, at the same time generating various interesting outdoor resting spaces.

商业裙房的形态在传统大型商场的基础上增加了室外连廊，并在商场顶上叠加了商业街模式，商业流线更为顺畅，业态丰富，以获取极佳的使用效率。这种新的裙房模式也提升了整体的商业价值，同时创造了多样的室外商业休憩空间。

Chapter Three

Suzhou Hong Leong City Center

Chapter Three

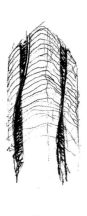

Apartment Low Zone (F3-F10)
塔楼1-公寓低区平面 (F3-F10)

Apartment Typical Plan (F10-F25)
塔楼2-公寓平面 (F10-F25)

Suzhou Hong Leong City Center

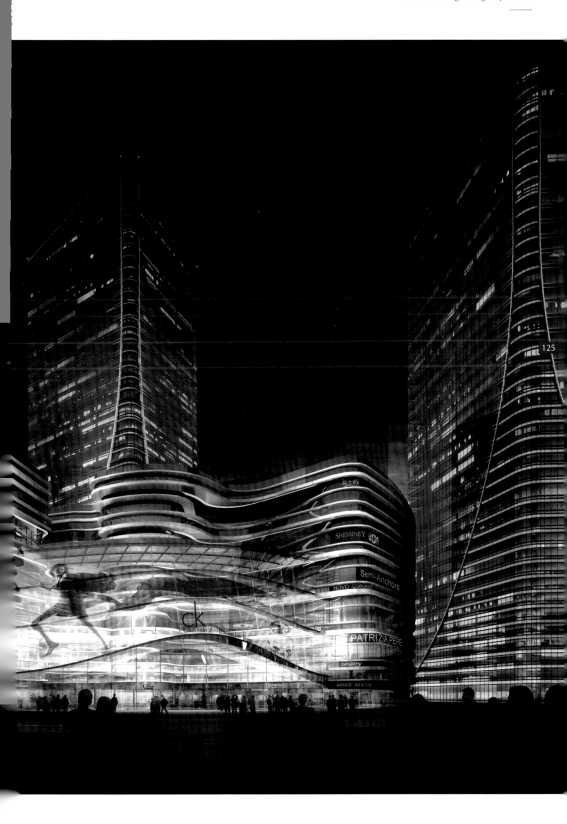

Chapter Three

Butterfly 蝴蝶

画舸春眠朝未足,梦为蝴蝶也寻花。

——《江行》(唐)鱼玄机

Travel on River by Yu Xuan Ji, Tang Dynasty

Delicate shapes, vibrant colors, flickering in a blooming field: a scene of butterflies can arouse thoughts of whimsical beauty and romantic ideals. From the ancient tale of Zhou Zhuang's dream, to the legend of Liang-Zhu's metamorphic love story, the butterfly has served as a motif of dramatic fantasy in Chinese literature and folklore.

The dancing figures of butterflies have often proved to be an elemental staple to the poetic and the picturesque. This project is committed to creating an eco-friendly, sustainable, symbiotic city; a city for "butterflies" to gather and dance.

The introduction of wetland, the water park and yacht clubs into the project, along with the development of a new architectural form, were dictated by the principle of people and nature interacting in harmony. The "butterfly" here is the nature's messenger, providing us guidance and endless inspiration.

蝴蝶，总能引起人们关于美好和浪漫的联想。纤巧的外形，绚烂的色彩，飞舞于烂漫山花之中，真是一番迷人的美景。在中国，自古有周庄梦蝶的典故，梁祝化蝶的传说，"蝴蝶"又多了一层梦幻的意义。

蝴蝶的舞姿总伴随着一番诗情画意。此项目致力于打造一座生态城，让城市如"蝴蝶"般的建筑聚集于此，翩翩起舞。

项目中，湿地、水上乐园、游艇俱乐部的引入，对新建筑形态的开拓，都是以人与自然的和谐互动为出发点的，而"蝴蝶"在这里是自然的信使，给我们以无限灵感和指引。

Chapter Three

Sino-Singapore Tianjin Eco-city Plot 8 &17 Planning
中新天津生态城开发区第8、17地块规划方案

Location: Tianjin, China
Type: Urban Planning
GFA: 326,500m²

项目位置：中国 天津
项目类型：城市规划
项目面积：326,500m²

Derived from nature, the main concept is inspired by a vivid image of a butterfly, with its enchanting patterns on its wings expanding across the sky. The architecture and landscape become infused into one fluid, continuous, wind-like form.

从自然中汲取灵感，借鉴蝴蝶的美妙形象，曼妙的曲线在基地上的生长延伸，整个项目犹如张开双翼的蝶。建筑和景观延续了曲线的柔和感和流动性，人造建筑与自然景观融合在一起，犹如流淌的河水，徜徉的风。

Sino-Singapore Tianjin Eco-city Plot 8 &17 Planning

Green Landscape 绿色景观
Wetland 湿地

Water 水景观
Wetland 湿地

Strategies to apply biological diversification and sustainability on the macro and micro scale are important so as to utilize available resources and reduce energy-related waste.

设计中注重生态平衡和可持续理念，从整体规划到建筑细节，都力求充分利用资源，减少能源消耗。

Chapter Three

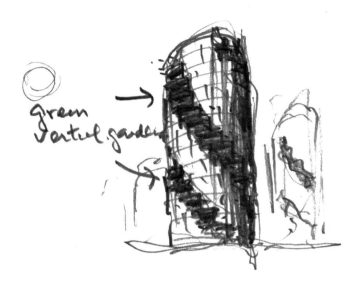

Sino-Singapore Tianjin Eco-city is located inside a protected ecological wetland on the northern district of Tanggu, Tianjin, with a close proximity to the bay areas. Due to the significance of the location and its potential impact to the surrounding ecological systems, it is crucial that the project protects the available ecological resources around the site.

中新天津生态城位于天津塘沽区北部湿地生态保护区，靠近入海口，毗邻城市生态隔离廊道。项目特殊的地理位置使其对周边的生态系统具有重要的意义，如何有效保护和利用资源，维护区域生态安全成为整个设计中很重要的课题。

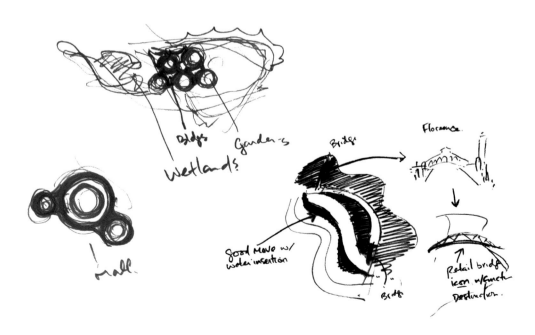

Sino-Singapore Tianjin Eco-city Plot 8 &17 Planning

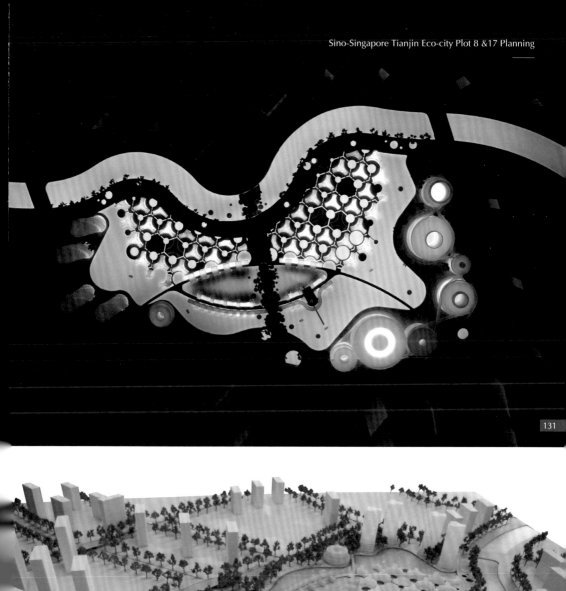

Chapter Three

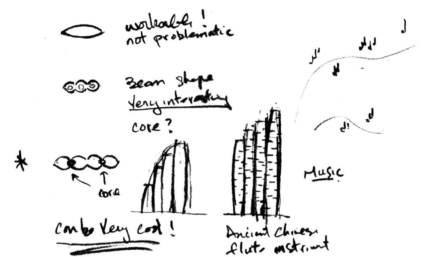

Sino-Singapore Tianjin Eco-city Plot 8 &17 Planning

The commercial zone is based on two concepts: uniformity and humanism. Uniformity as to integrate various styles into a unified whole and humanism as to provide an ideal space for entertainment and shopping. As the tallest structure planned along the river bank, the architecture of the residential towers must appear unique and iconic.

项目的商业区在设计上追求"整体性"和"人性化"的特点。所谓"整体性"是指建筑群相对拥有统一的风格。所谓"人性化"是指其在群体结构上为人们提供一流的购物环境和休闲体验。高层住宅塔楼沿河而布,作为园区的至高点,建筑形象要求创新独特,具有标志性。

Shopping Mall Massing
购物中心体量

Sino-Singapore Tianjin Eco-city Plot 8 &17 Planning

Club House
会所

The residential towers are organized along the river bank. As the highest point inside the ecological complex, their architectural appearance needs to be unique and iconic. Adapting features from the traditional courtyard housing, the low-density residential units are configured as a circular space to maximize day light. Local traditions are re-interpreted to create a healthy contemporary lifestyle.

The large balconies are set to reinvent a luxurious city lifestyle. Additionally, the three-dimensional landscape system creates a multi-functional system to increase the garden's overall ecological performance.

低密度住宅创造了一种圆形院落空间，既沿袭了传统住宅院落空间的优点，又解决了建筑朝向问题，传统温馨的聚居方式在此以更为健康现代的面貌得以呈现。

大面积跃层阳台开创了新型现代奢华的生活方式，同时，立体化的景观体提升了建筑整体的生态功能。

Chapter Three

Dragon 1　　盘龙

龙行踏绛气，天半语相闻。混沌疑初判，洪荒若始分。

——《奉和登骊山应制》（唐）阎朝隐
Chinese Poem about Dragon by Yan Chaoyin, Tang Dynasty

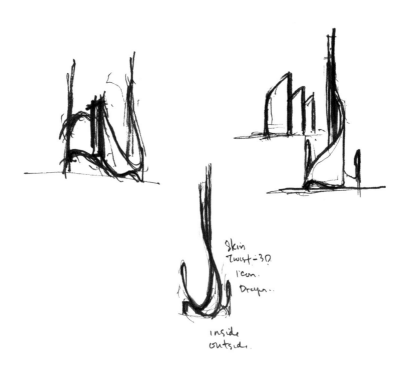

The dragon is a supreme entity in Chinese mythology. In ancient times, it was believed that dragons possessed the power to alter the weather by creating clouds and rain. This godly being became the symbol of the imperial house in feudal society and an international icon for Chinese work ethic in recent times.

This project is located in the historic capital of Nanjing. Our objective was to develop an architectural vocabulary that could rouse national strength. The inspiration for design came from the traditional dragon sculpture; the main structure spiraling up into the sky in elegance, denoting the instantaneous power as the dragon prepares to soar.

龙在中国的神话传说中,有至高无上的地位。在远古时代,龙是一种善变化、兴云雨、利万物的神兽。在漫长的封建社会,龙成为皇权的象征。在近代,龙更成为整个华夏民族的代表,为世界所认知。

项目位于古都南京,我们希望可以寻找一种建筑语汇,唤起民族内心的某种力量。设计从传统的盘龙雕塑中汲取灵感,舒展俊逸的建筑主体盘旋而上,直冲云霄,展现了飞龙在天前蓄势待发的一瞬间所迸发出的巨大能量。

Chapter Three

Nanjing Hunan Road Plaza
南京湖南路广场

Location: Nanjing, China
Type: Mixed-use
GFA: 635,700m²

项目位置：中国 南京
项目类型：综合体
项目面积：635,700m²

Located at the northwestern district inside Nanjing City, Hunan Road is a popular street known for its lively blend of retail, food services, local markets, leisure and tourism. Situated on a significant urban node, the project's main tower is 485-meter tall, envisioned to become a new icon for the city.

The massing incorporates an elevated pathway that wraps along the building's exterior plane, a form derived from the image of a dragon. An illuminated glass box at the top of the building embodies the head of the dragon, as the tower's remaining structure represents the body and tail of the dragon. Green spaces are contained within the artificial skin, conveying a reciprocal relationship between nature and human.

湖南路位于南京市城区西北部，是古城南京集商贸、饮食服务、文化娱乐和休闲旅游为一体的著名商业街。项目位于湖南路西端，是整座城市的一个重要节点，主塔楼达到485米，建成后将成为南京的新地标。

建筑设计以龙为寓意，在体量上塑造出一条沿塔楼外围蜿蜒盘旋而上的绿色景观带，宛若一条盘龙围绕塔楼向上攀升。顶端发光玻璃天棚为龙头，塔楼主体部分为龙身，裙房部分为龙尾。立体的自然景观带被人工建筑表皮所包裹，在大气统一的外观形象下体现的是人与自然的和谐互动。

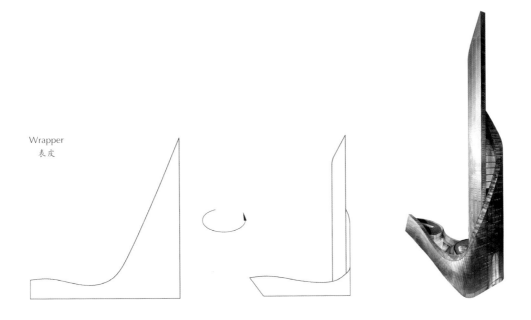

Wrapper
表皮

Chapter Three

The design emphasis is placed on creating a unique urban public space. The main entrance leads to a commercial plaza that contains escalators and a grand staircase, developing a spatial complexity that enhances the commercial atmosphere.

The space between the west side of the apartment and the urban park contains a recreational, water-themed urban plaza. In the summer, it can perform as an amphitheater and, in the winter, it can transform into an ice-skating rink.

设计注重城市公共空间的营造。商场主入口面向城市广场，大台阶和自动扶梯增加空间层次和商业氛围。

公寓楼西侧与城市公园相连的空间布置休闲水上城市广场，夏天可作露天剧场，冬天可作冰场。

Nanjing Hunan Road Plaza

Chapter Three

Nanjing Hunan Road Plaza

广场绿化 屋顶绿化 道路绿化 湖滨绿化

The design for green landscapes is the highlight for this project.

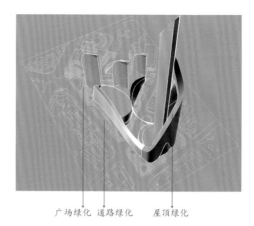

广场绿化 道路绿化 屋顶绿化

立体化的绿色景观空间设计是项目的亮点之一。

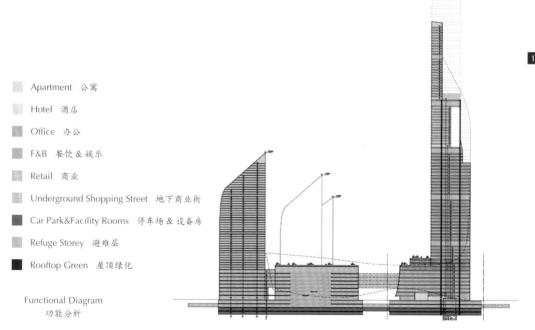

- Apartment 公寓
- Hotel 酒店
- Office 办公
- F&B 餐饮 & 娱乐
- Retail 商业
- Underground Shopping Street 地下商业街
- Car Park & Facility Rooms 停车场 & 设备房
- Refuge Storey 避难层
- Rooftop Green 屋顶绿化

Functional Diagram
功能分析

As a large scale urban complex, the Hunan Road Plaza consists of apartments, hotels, offices, entertainment, and commercial related functions. The main road divides the site into two zones; a bridge-like structure links the zones together to maintain the continuation of the site.

湖南路广场作为一个大型的城市综合开发项目，主要包括公寓、酒店、办公、餐饮娱乐、商业等功能。基地被城市道路分为两个地块，为保证建筑形象的完整性和连续性，用过街楼方式连接两地块。

Chapter Three

Nanjing Hunan Road Plaza

Chapter Three

Dragon2　蛟龙

扶河汉，触华嵩，普厥施，收成功，骑元气，游太空。

——《九龙图》题诗　（南宋）　陈容

Kowloon Map by Chen Rong, Song Dynasty

Hengqin International Financial Center

Adjacent to mountains, seas, and conveniently surrounded by four metropolises: Hong Kong, Macau, Shenzhen, Zhuhai, the Hengqin New District possesses powerful momentum for future development. The design is an attempt at embodying a convergence of energy in architectural form. The traditional totem of the dragon emerging from the sea, symbolic of new life breaking through the barriers of limit, flourishing endlessly, is a very fitting image for the project.

Inspired by the shapes of the "Kowloon Map", we seek to express the quality of speed, power, and vitality through architecture in an atmosphere of rising clouds. The location, sheer volume of the project, its programs, provide an excellent platform for design. This is an example of the contemporary Chinese architecture that reflects the current growth of country as well as the hopes and dreams for the future.

横琴新区背山面海，被香港、澳门、深圳、珠海四大都市所环绕，无比优越的地理位置提供了源源不断的发展动力。如何通过一个建筑来传递这种震撼人心的生命力？"蛟龙出海"这一中国的传统图腾，寓意新生的力量冲破困难蓬勃发展，似乎非常贴合。
造型的灵感来源于《九龙图》，我们尝试用建筑的方式展现无穷变化的神力，粗犷迅驰的线条，威猛布空的气势，云气升腾的意境。项目的位置，体量，功能定位都为这种尝试提供了极佳的平台。这是一个属于当下中国的建筑，映照出这个国家所正在发生的那些与梦想相关的事情。

Chapter Three

Hengqin International Financial Center
Co-designer: Keith Griffiths
横琴国际金融中心大厦

Location : Hengqin, China   项目位置 : 中国 横琴
Type: Mixed-use   项目类型: 综合体
GFA: 138,158m²   项目面积: 138,158㎡

The Pearl River Delta is a financial center in southern China. Hengqin Island is situated south of Zhuhai, connected to Macau to the east while overlooking Shenzhen and Hong Kong across water. This once quiet fishing island grew into an important economic zone due to its convenient location.

珠江三角洲是中国南部的经济和金融中心。横琴岛位于珠海南部，东面与澳门一水相隔，与深圳、香港隔海向望。这个曾经宁静的渔业小岛因其优越的地理位置逐渐成为世人瞩目的焦点。

Hengqin International Financial Center

Chapter Three

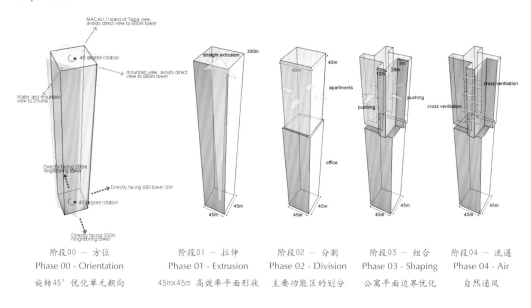

The project is situated at the Central Business District, the north end of Hengqin district. It is a large, urban mixed-used complex including the programs of office, apartment and expo.

项目用地位于十字门中央商务区横琴片区北端,功能定位为集办公、公寓、会展为一体的大型城市综合体。

Hengqin International Financial Center

 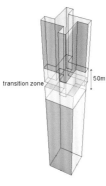 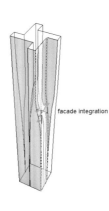 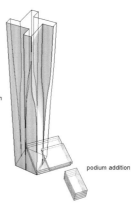 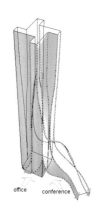

阶段05 — 位置
Phase 05 - Basic
提升公寓品质

阶段06 — 转换
Phase 06 - Transition
设置转换层

阶段07 — 融合
Phase 07 - Merging
连接两层表皮

阶段08 — 裙楼
Phase 08 - Podium
植入裙楼

阶段09 — 统一
Phase 09 - Draping
裙塔合一

Office Plan-Low Zone
办公低区平面

Office Plan-High Zone
办公高区平面

Apartment Upper Zone Plan
公寓高区平面

Chapter Three

The building spirally rising from podium looks like four towers integrating into one. The canopy of office entrance is extended from tower facade which resembles a curtain lifted by the wind; The retail entrance adopts curved flowing line to attract pedestrians; Both the grand atrium of the exhibition hall and the dynamic showcase of flagship store help to create an unique spatial experience and impressive architectural form.

建筑的四栋塔楼由裙楼扭转向上，仿佛合为一栋。办公入口雨篷曲面与塔楼立面自然连接，酷似被风吹起的垂帘；商业入口则采用弧面流动的线条吸引人流；会展大堂通高的玻璃中庭，旗舰店橱窗的几何拧转都创造出建筑裙房独一无二的空间感受和形体特征。

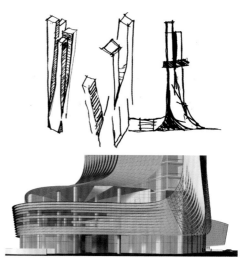
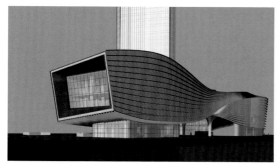
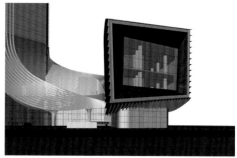

Hengqin International Financial Center

High-end business apartment on the upper level of the tower is developed into windmill-shape which optimizes not only ventilation conditions but also view advantage for the house.

塔楼上部的高端商务公寓呈风车型布局，既最大限度的为户型提供通风条件，又增加了景观面。

Chapter Three

The facades are composed of mainly metal plates, and two different curtain wall glass to highlight its unique architecture form.

The podium facade are designed with parametric surfaces, which can be divided as 1.5mX1.5m triangular elements. The materials applied to each face often based on the functional requirements of the corresponding spaces on the interior, thus creating a powerful dynamic between the inside and the outside, at the same time fulfilling a compelling and unique architectural gesture.

立面主要通过金属板和两种幕墙玻璃材质来表现独特的建筑形态。

裙房部分通过曲面的参数化设计，采用1.5mX1.5m的三角形单元对外表皮进行划分，然后依据室内功能要求和外立面效果选择不同的材料对单元进行填充，以达到独特而夺目的建筑形象。

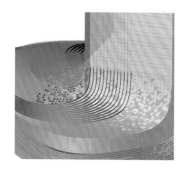
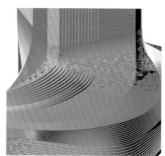
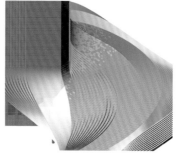
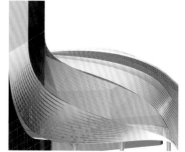

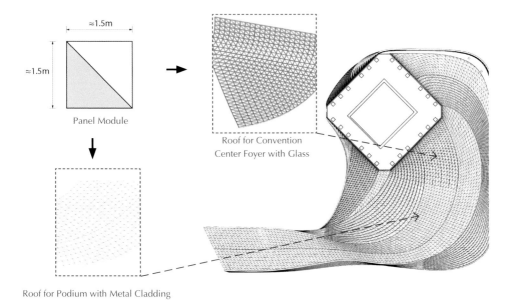

≈1.5m
≈1.5m

Panel Module

Roof for Convention Center Foyer with Glass

Roof for Podium with Metal Cladding

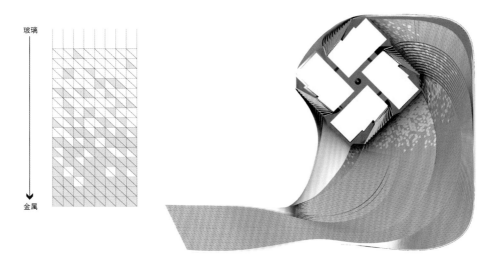

玻璃

金属

Chapter Three

Hengqin International Financial Center

Chapter Three

The interweaving-shaped canopy of convention entrance is designed into an integrated form with its supporting columns resembling intersecting branches under a banyan tree.

将会展中心入口处的顶棚与柱子统一设计，利用有序交织的设计手法，营造出酷似榕树树枝纵横交错的感觉。

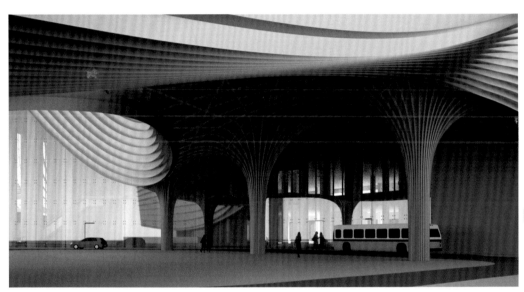

Chapter Four

CHAPTER FOUR 第四章

Tales of Humanity

The following pages detail the most exciting moments of this architectural journey.

The ancient Greeks told us through the aphorism "know thyself" engraved on Temple of Apollo at Delphi, that curiosity and self-exploration will become the driving force for human beings. As we approach the end of the book, we would like to capture the valued moments of this journey in our own language and share it with our readers.

人类的故事

终于到了这个章节,也许有些自大,这是故事最精彩的部分。

正如刻在德尔斐的阿波罗神庙的那句箴言:认识你自己。人对于自身的好奇和探究永无止境,故事还在继续。在这本故事集尾声将近时,我们尝试用一种特别的语言撷取这个动人旅程中的点滴,与您分享。

Tales of Humanity

Chapter Four

African Dance 非洲舞

Do you hear the beat of my African drum?

It goes at a steady beat.

Do you see my African man?

He dances as if he was the wind god dancing.

— On the Clouds

《云端》

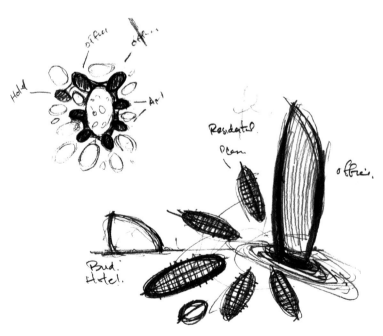

Angola Mixed-use Commercial Center

Tahiti is Paul Gauguin's paradise on earth. Under the sun, within the sea, it is the birthplace of distinctive lines, tenacious volumes and vibrant colors. In his paintings, Gauguin opened the window to these mysterious, animalistic, vivacious Tahitian people, whose primitive, rhythmic dances shocked, bewildered, and inspired a modern society desensitized by daily monotony, blind materialism and cultural "refinement". We cannot continue to suppress our instincts; we must embrace and celebrate the fiery enthusiasm and admiration for life that makes us humans.

The African continent is a magical land, one closest to the sun, where mankind can trace back to the absolute origin of its evolutionary journey. This was the subject raised in Gauguin's masterpiece titled: "Where did we come from? Who are we? Where are we going?" From time to time, we must glance backwards to our primordial past and listen to the sounds of nature to find inspiration.

It was an exciting experience, to be involved in this African project. Struck with wonder and respect, we set out to complete the ceremonial ritual through architecture. In an act to immortalize the vitality, purity, mystery and majesty of the native dance, we will narrate a story of celebration, unity, history and honor.

塔希提岛在保罗·高更眼中是人间天堂。阳光之下，碧海之中，孕育出明晰的线条、结实的体量、强烈的色彩。高更笔下那些神秘、野性、炙热的塔希提人，跳着富于节奏感的原始之舞，震惊了无数被文明麻痹了的现代人。我们无法继续无视这种本能，我们的内心激荡着的和祖先一样的对自然和生命的崇敬。

非洲大陆，是离太阳最近的神奇土地，人类能够在那里找到自己的起点。正如高更的名画《我们从哪里来？我们是谁？我们往哪里去？》，无论在什么时代，总会有声音代表人类发出这样的疑问。答案也许就在原点，时不时的回望一下我们的过去，聆听一下自然的声音，也许就能寻到某种启示。

参与非洲的项目是令人激动的一种经历。我们满怀敬意，希望通过建筑来完成一种仪式性的祭奠。项目灵感来源于非洲舞蹈，尝试用建筑群来重现原始舞蹈的活力、纯净、神秘和雄壮，同时也在诉说一个关于庆典、团结、历史和荣誉的故事。

Chapter Four

Angola Mixed-use Commercial Center
安哥拉综合商业中心

Location: Luanda, Angola
Type: Mixed-use
GFA: 426,059m²

项目位置：安哥拉 卢安达
项目类型：综合体
项目面积：426,059m²

Angola Mixed-use Commercial Center

As the capital and largest city on the coast of Angola, Luanda is a chief seaport and administrative center, currently undergoing a major regeneration with large developments underway that will significantly alter the cityscape.

Located at the heart of Luanda, the inspiration for the planning of the Angola Mixed-use Commercial Center comes from an image of an African tribal dance around a body of fire. Eight different towers are connected by the podium, creating a unified and ceremonial identity. The aggregation of these buildings evokes a mental image of an African landscape and a traditional African mask.

卢安达是安哥拉的首都和行政中心，也是安哥拉最大的城市，毗邻大西洋，是非洲的主要海港。随着地区的复兴，卢安达渐渐成为世界的新热点，很多大型地产开发项目在城市的各个角落开展，这个非洲城市正在发生巨变。

安哥拉综合商业中心位于卢安达中心城区，规划从非洲部落围绕篝火翩翩起舞的画面中汲取灵感，八个不同功能的塔楼被中央裙房环接起来，构成一个庆典、团结、历史和荣誉彰显的整体形象。新颖的建筑群体形象唤起人们关于非洲这片土地的遐想，还有海边的贝壳或古老的面具……

The Angola Mixed-use Commercial Complex is a large scale, multi-functional complex. Each of the eight towers includes office, hotel, and residential functions while the podium includes a high-end commercial center and parking lot. Each program performs separately, yet cohesively, as an entity.

安哥拉综合商业中心是一个大型城市综合开发项目，裙房内主要为高端商业中心和停车库，八栋塔楼内布置有办公楼、酒店和住宅等功能。各功能间既相互独立，又有关联性，成为一个有机的整体。

Chapter Four

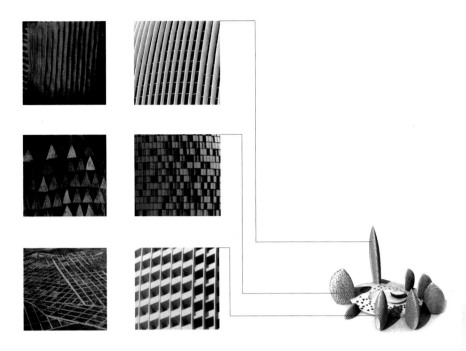

The design patterns applied to the facade come from traditional African crafts and contextual materials, each corresponding to unique functions.

The landscape design is integrated into the architecture. The roof of the podium uses both hard and soft pavements to depict traditional African craftsmanship, as well as creating a unique facade.

Pattern of Roof Garden
屋顶花园

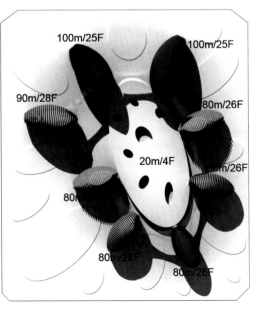

Height Analysis
高度分析

Angola Mixed-use Commercial Center

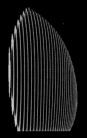

Administration and Information Building
行政信息楼

Administration and Information Building
行政信息楼

建筑立面的灵感来源于非洲的手工艺品和织物材料，不同的肌理对应不同的建筑功能。

景观设计与建筑形象相协调，裙房的屋顶平台以软硬铺装相结合，呈现原始工艺品般的美妙纹理，创造了独特的第五立面。

Chapter Four

Chapter Four

Lover 情人

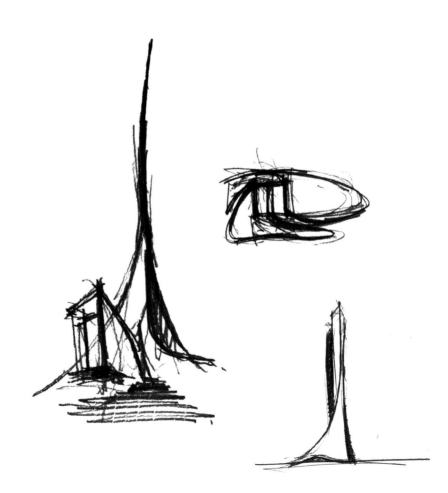

Nanjing Hexi Olympic Plaza

Elegant curves of bodies are intertwined, displaying an elemental dignity of humanity. Rodin's famed sculpture, "The Kiss", solidifies an intrinsic moment of contact between man and woman. The kiss is an immaculate profession of pure love, a primal emotion and an recurrent theme in countless stories. The process of pursuing love and joy is always accompanied by pain and struggle. It's these moments of passion and desire, sin and punishment that highlight the vitality of life.

Can a building generate this type of sensation? The development of modern construction technology has given architecture endless possibilities. Through architecture, we attempt to simulate an emotion, a moment of grace, rather than a mere machine for living. Architecture should be able to express human sentiments; it is a medium for communication between the architect, the city and the times.

罗丹的《吻》，捕捉了纯洁男女的最初接触，细腻优雅的肌体和姿态，呈现出人性的尊严，凝固了最动人心弦的一瞬。吻，常常有某种圣洁的意味，与爱有关，是世间不朽的主题。人们追求爱和欢乐的过程中，总是伴随痛苦和挣扎，在爱与欲、罪与罚中，爆发出感人的生命力。

建筑是否也能带来这样的感动？现代工程技术的发展赋予了建筑更多的可能性，我们尝试塑造一个优雅感人的建筑形象。不再是城市中冷漠的居住机器，建筑应该可以表达更多的感悟，成为建筑师与城市、与时代沟通的媒介。

Chapter Four

Nanjing Hexi Olympic Plaza
南京河西奥体新城广场

Location: Nanjing, China
Type: Mixed-use
GFA: 385,600m²

项目位置：中国 南京
项目类型：综合体
项目面积：385,600m²

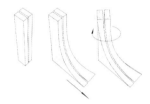

The Nanjing Hexi Olympic Plaza is located on the west of the CBD, northeast of Nanjing International Convention Center and southwest of Olympic Center. Standing 380 meters tall, the tower will become an important landmark within the West Nanjing River Olympic City.

The plaza is designed to express two sculptural figures "kissing" as it intersects and twists when elevating from the ground. Each figure is corresponded to its materiality, with the black tower as the male and the white tower as the female. This creates a differentiation and contrast between the towers, embodying the concept of Yin and Yang.

The building consists of a series of intricate lines and planes to emulate a gradual change of angles. Accentuated in form and lighting, the project solidifies a moment of a passionate dance between man and woman, in a sensuous twist of the gown and the bodies.

南京河西奥体新城广场位于河西CBD中央商务核心地区西侧，南京国际会议中心东北角，奥体中心西南侧。塔楼设计高度为380米，建成后将成为河西奥体新城的重要组成元素。

广场造型动感，设计独特。灵感源于雕塑"吻"，创造出双塔形式，随着上升的高度而交叉扭转。两座塔楼相互对照，不同风格的建筑外墙保持自身的独立性，白塔象征"女性"，黑塔象征"男性"，突出了黑白的对比，阴阳的结合。

整座建筑有着复杂的曲线和层次，力图通过多角度变化自身形态。建筑表面在自然环境和人工光线的作用下更突出扭转的效果，呈现出旗袍般的柔美和双人舞般的默契。

Chapter Four

Nanjing Hexi Olympic Plaza combines architectural elements of the green sunken courtyard, green pavement, green roofs and vertical gardens, forming a unique architectural experience.

The stepping green roofs offer visitors an open space and an interior green space.

The suspended space between the two towers consists of a series of double height green platforms, enhancing the overall spatial quality.

The sunken courtyard in front of the main entrance attracts public circulation from the ground and connects to the nearby metro station. It also integrates nature into commercial spaces to maximize its commercial value. The steps provide an open space for public interactions and direct circulation onto the roof of the commercial center.

The circular commercial courtyard - expands the views for shoppers. The unique change in levels provides the public with a place to socialize, while introducing elements of nature onto the rooftop.

南京河西奥体新城广场由下沉广场绿化、地面绿化、裙房屋顶绿化和塔楼垂直绿化等元素构成连续的立体景观体系，丰富的自然元素穿插于建筑的扭合之间。

退台式屋顶绿化，为购物者提供了更多的开敞空间和室内绿色环境。

两塔间的挑空空间内设计了错层式空中绿化平台，提供了更多开敞空间。

主入口处的下沉广场吸引了地面人流。与地铁车站衔接，自然的将人流引入了商业中心，最大限度的提高了地块的商业价值。主入口处的台阶为人们提供了聚会与交流空间，并自然地将人们引入了商业中心的顶层。

商业中心内部圆形中庭，为购物者提供了开敞的视觉空间。弧面天窗，以及卵石形的多功能厅，为中庭创造出美妙、奇特的空间，将自然的元素引入顶层。

Nanjing Hexi Olympic Plaza

Chapter Four

City Files I – Modern Church 城市档案一 —— 都市教堂

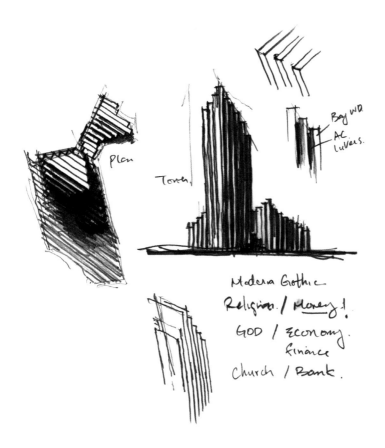

Religion played an influential role in the making of European culture. For many centuries, the Catholic church stood as the most important structure in the city. In the novel "Notre Dame de Paris", a story fueled by determinism, social dynamics and the perpetual struggle between the pure and the sinister; almost every major event taking place inside the cathedral, or witnessed by another character standing within or atop the cathedral. As the stars of the story perish, the Notre Dame continues to stand in its enduring majesty, silently witnessing the coming and going of humanity's histories and tales.

Gothic architecture is characterized by rising lines to create magnificent exteriors and spacious interiors, contributing to the creation of an imposing religious atmosphere. The architectural intention of "touching the sky" reflects the religious desire of "extending towards the heaven." In the long course of history, the Gothic cathedral is the perhaps the most convincing attempt to unify architecture and spiritual ideals.

In modern society, consumer culture has become an inseparable part of urban living. City centers are mostly dominated by shops and malls. Fervent materialism has become ubiquitous. Consumers today share a similar enthusiasm as devout believers had in the past, as shopping centers and churches seem to have a certain urban architectural equivalence. "Modern Church" was the concept we used to convey this idea throughout the project. Vertical lines and an elegant glass curtain wall embody our contemporary interpretation of this spirit.

宗教曾是欧洲人生活中的重要组成部分，教堂成为城市中的一处重要场所。在《巴黎圣母院》中，我们读到了一座哥特建筑中所上演的爱情经典。故事中的主角已随风而逝，巴黎圣母院依旧以其威严和不朽默默见证着芸芸众生。

哥特式建筑以直升线条、雄伟的外观和教堂内空阔空间，渲染了浓厚的宗教气氛。"触摸天空"的意向象征了"升向天堂"的宗教理想。在漫漫历史长河中，哥特式教堂是达成建筑形象和精神理想的极致统一的典范。

探索现代都市人的生活，"消费"成为我们生活中很重要的部分。城市中心往往被重要的消费场所占据。对物质的崇尚不亚于曾经信徒对于主的虔诚，这种意义上而言，商场与教堂有某种城市功能上的对等。"都市教堂"，是我们想通过这个项目传达的概念。竖向的线条，简洁的玻璃幕墙，是我们对当代精神的解读。

Chapter Four

Huai'an Suning Plaza

淮安苏宁电器广场

Location: Huai'an, China
Type: Mixed-use
GFA: 87,329m²

项目位置：中国 淮安
项目类型：综合体
项目面积：87,329m²

Huai'an is a historically significant site, located on the border that divides northern and southern China. Today, Huai'an is undergoing a transformation to become a lively and sustainable urban center in the northern region of the Yangtze River.

The Suning plaza is located in a prosperous district within the city, a highly valued lot which connects to the Huaihai Plaza from the south. Due to the site's complexity and irregularity, the challenge of the project is to efficiently utilize limited space, and to resolve the relationship between Huai'an Suning Plaza and the surrounding buildings.

淮安位于中国地理南北分界线上，历史悠久，人杰地灵。今日淮安正在建设成为长江三角洲北部地区重要的中心城市和具有绿色生态特色的宜居城市。

苏宁广场位于淮安最繁华的城市地段，南接淮海广场这一重要的城市广场，基地形状曲折不规整，如何有效利用基地，并处理好建筑与淮海广场以及其他周边建筑道路的关系成为设计的一个重要课题。

To address the limited site boundary, the design combines a series of 45 degree cuts which create an array of rhythmic lines that are both contemporary and elegant. The folded architectural form creates a vibrant street front leading to diverse, high-valued corner spaces, illustrative of the term "golden angles and golden seams." This approach offers visibility from multiple directions to the dispersed commercial signs.

45度角切片的形式将塔楼的正立面调整到朝向淮海广场路口，巧妙解决了基地形状不规则造成的困难，实现"华丽转身"。节奏清晰、挺拔优雅的垂直线条既高贵典雅又清新现代。建筑折线形的边界创造出更充足的商业临街面，形成更多具有很高商业价值的转角空间，所谓"金角金边"，并有利于在多个方向上加强商业标志的可视性。

Chapter Four

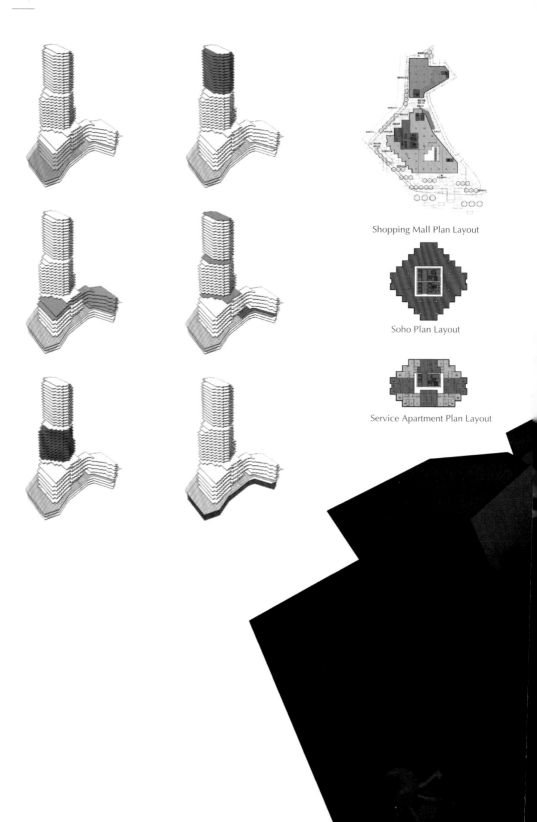

Huai'an Suning Plaza

Chapter Four

City Files II – History 城市档案二 —— 历史

滚滚长江东逝水,浪花淘尽英雄。是非成败转头空,青山依旧在,几度夕阳红。

——《临江仙》(明)杨慎
Lin Jiang Xian By Yang Shen, Ming Dynasty

Shanghai became an open city in 1843 and became a prominent city with historical importance in China. In less than 200 years, this city has undergone massive transformations. This process has been documented by countless people. In many of their eyes, this river accompanied the city through all of its tumultuous past. It had been clean, it had been muddy. Perhaps its shade will never return to the translucence of water, but very much like the city itself, it gives its inhabitants a chance to trace their history and roots.

Architectural design at times, is to create something out of nothing. The government approves the sparse, clean, regulated parcel, as long as it complies with the limits of the redline and setbacks. The height limit remains an abstraction, the designer is full access to the third dimension. Yet, there is no point in time and space which stands in complete isolation. The designer must express these implications of history, evolution and continuity.

Located on the north bank of Suzhou River, the project helps to reshape the waterfront, and integrates the Joy City Community into the transforming urban context.

上海自1843年开埠以来，至今170年历史，在泱泱的华夏历史上实在有些数不上。可正是这座城市在不到200年内历经巨变，有无数的人用不同的方式来记录她的各个瞬间。在众多的见证者中，有这样一条河流，陪伴着这座城市从历史深处悠悠走来，她清澈过也浑浊过，她也许再也回不去那属于水的透亮，但她就如这座城的历史，给生活在这里的人一个溯源寻根的机会。

文脉，是一个需要诚心对待的问题。建筑设计以某种角度而言是一种无中生有的创造，政府总是能做出一块空荡而平整的场地，似乎只要满足了对那根抽象的红线的退界要求和那个抽象的限制高度，设计者似乎就能在抽象的三维空间内为所欲为了。事实并非如此，没有时空是孤立的，一个设计者必须把时空的连续性纳入他的思考中。

当我们有机会参与苏州河的变迁，当我们的作品将矗立在苏州河边，如何真正地实践建筑与城市文脉的融合成为我们必须面对的一个挑战。

Chapter Four

COFCO Shanghai Joy City Phase 2

COFCO Shanghai Joy City Phase 2
上海大悦城项目二期

Location : Shanghai, China
Type: Mixed-use
GFA: 108,930m²

项目位置：中国 上海
项目类型：综合体
项目面积：108,930m²

Joy City is located at the core of Suhei Bay and linked to prime commercial districts. Phase I retail mall, after construction, stand as a unique commercial landmark. Phase II aims to integrate further into the urban fabric, at the same time rejuvenate urban life with vibrant activities as a dynamic mix-use complex.

大悦城位于上海苏河湾核心区域，与重要商圈衔接。已建成的一期商场因其鲜明的个性定位，成为独具时尚魅力的商业地标。项目二期建设旨在与城市文脉相结合，同时协同一期构建一个富于活力的城市综合体。

Chapter Four

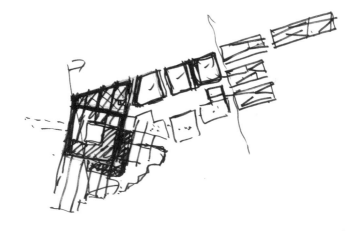

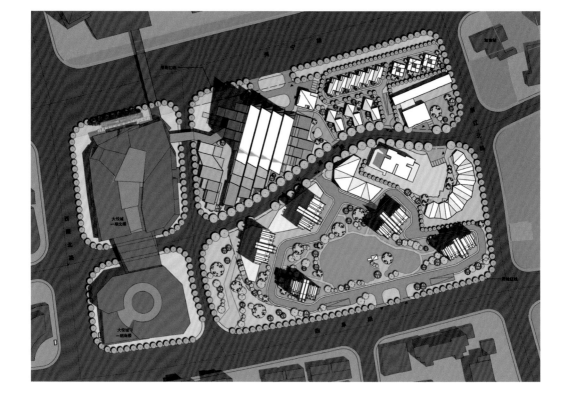

COFCO Shanghai Joy City Phase 2

The plot of Phase II is at the east of Phase I, its functions include office, commercial and apartments. It includes a 300m tall super high-rise tower. The land is adjacent to the Suzhou River; facing the east and south are fantastic views and abundant resources.

There are numerous historical sites to the northern part of the parcel. The design explores the various ways to integrate new architecture on top of the existing, and to express their dialogue.

二期用地位于一期用地东侧，规划功能以办公、商业、公寓为主。其中包含一栋300米高地标性超高层塔楼。地块毗邻苏州河，向东向南均有充沛的湾景资源。

北面地块内有多个历史保留建筑，新建建筑与保留建筑的和谐共存成为项目设计的主要出发点。

办公塔楼 Office Tower
商业 Retail
酒店式公寓 Service Apartment
住宅配套商业 Commercial Facilities
住宅 Residential
幼儿园 Kindergarten

Chapter Four

As a new regional landmark, the main tower is a tribute to the history of Shanghai and the Suzhou River. The office towers portray a series of overlapping, golden sails, carrying the essence of Eastern and Western civilization.

作为地区新地标的主塔楼,将是一座向上海和苏州河历史致敬的作品。那些曾经在黄浦江和苏州河上扬起的风帆,承载了中西方文明的精华,造就了上海当年和今日的繁荣。本项目的核心办公塔楼仿佛重叠的金色风帆,将永驻于苏州河畔。

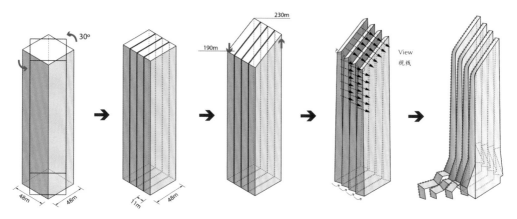

Phase 01 - Orientation
阶段01 — 转向

Phase 02 - Division
阶段02 — 分割

Phase 03 - Pitched Roof
阶段03 — 尖顶

Phase 04 - Movement
阶段04 — 错动

Phase 05 - Extension
阶段05 — 延展

COFCO Shanghai Joy City Phase 2

Chapter Four

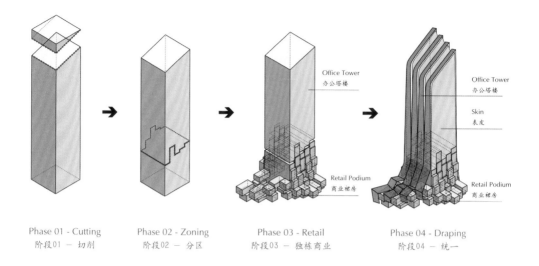

| Phase 01 - Cutting | Phase 02 - Zoning | Phase 03 - Retail | Phase 04 - Draping |
| 阶段01 — 切削 | 阶段02 — 分区 | 阶段03 — 独栋商业 | 阶段04 — 统一 |

The layout of the commercial streets surrounding the main tower, are derived from a series of staggering, overlapping blocks, forming an intimate human scale and evoking a sense of the "garden house" lifestyle, which is unique to Shanghai. A canopy extend down from the tower facade, serving both as a shade for the retail street and connecting the high-rise office and the commercial podium into one form.

围绕主塔楼布局的风情商业街由一系列重叠交错的体块所构成，形成亲切的人性化尺度，同时唤起人们对上海独有的"花园洋房"生活模式的回忆。由塔楼侧面延伸下来的幕墙所形成的天幕，既为商业街遮阴，也将高层办公与商业裙楼组织为一个整体。

COFCO Shanghai Joy City Phase 2

Chapter Four

COFCO Shanghai Joy City Phase 2

  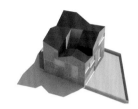

The low-rise apartments and the kindergartens employ a similar spatial arrangement as the architecture of Old Shanghai, incorporating a modern architectural vocabulary to a desirable urban lifestyle.

基地内的低层公寓和幼儿园的设计借鉴了上海旧城区核心地带的低密度居住社区的空间结构，试图用现代的建筑语汇重塑曾经令人向往的生活方式。

Chapter Four

City Files III - Hakka Walled Village　　城市档案三 —— 土楼

福建永安县贼邓惠铨、邓兴祖、谢大髻等,于嘉靖三十八年聚党四千人,占据大、小淘水陆要道,筑二土楼,凿池竖栅自固,且与龙岩贼廖选势成犄角。

——《重修虔台志》

Wuxi Metro Mixed-use Project

In the 1907s publication *The Production of Space*, French urban theorist Henri Lefebvre contends that the notion of space extends beyond volume and geometry. There are two distinct levels of spatial practices: absolute, physical spaces, or natural spaces, as well as socially produced spaces, or social spaces. Lefebvre derives that all spaces are social products, or construction of values and experiences, thereby allowing for a reproduction of society itself.

Through the observation of everyday life, other theorists have constructed the concept of the spatial trinity in urban planning, composed of "physical space", "spiritual space" and "social space."

The form of the Hakka Villages, or the Tulou, is governed by a series of concentric circles of varying heights and scales. Enclosed in towering walls, the village is a gated megablock, containing radial units connected by circular corridors. These rings are situated smoothly into the organic topography of natural hills. It is a totem of "physical space", as well as a self-sustaining community characterized by the traditional clan-styled social relations. Lifestyle, architecture and the inhabitant become interdependent factors dictating the evolution of one another.

While we employ contemporary architectural vocabularies to reshape traditional, we retain the values of physical and spiritual spatial practices. Design is an exploration of our origins, of the Garden of Eden, of Sumeru, of the heavenly realms.

20世纪70年代法国城市社会学家列斐伏尔提出"空间生产"的概念：空间不是通常的几何学与传统地理学的概念，而是一个社会关系的重组与社会秩序实践性的建构过程，空间性是社会的产物，是价值和实践的构架，关系到社会自身的再生产。

思想大师从对日常生活细致入微的观察中完成了城市空间理论的建构，即"物理空间"、"精神空间"、"社会空间"三位一体。

"土楼"，四周高墙耸立，多环同心圆楼外高内低，楼中有楼，环环相套，以天井相隔，以廊道相通，在卫星图上它们以几乎完美的几何形态凸显于自然丘陵之中，成为一个重要的"物理空间"图腾。与之相关的是宗族式的社会关系和传统的伦理精神。生活方式、建筑、人三者在这个圆环中相互依存，自在自观。

我们总是说通过现代的建筑语汇来重塑传统的空间，这里所说的也许不仅仅是物理上的空间，更多的是心灵上的。我们一次次地探寻回到原点的路，那里是伊甸园，是须弥山，是极乐天园。

Chapter Four

Wuxi Metro Mixed-use Project
无锡地铁综合体项目

Location : Wuxi, China
Type: Mixed-use
GFA: 200,650m²

项目位置:中国 无锡
项目类型:综合体
项目面积:200,650m²

The project is located in the Wuxi Coastal region, north of the Grand Canal, south facing Taihu Lake. The parcel is situated at an advantageous traffic node, containing the intersection of metro lines 3 and 4.

The circular geometry is a response to the complexity and multiplicity of the surrounding urban fabric. Four entry points introduce pedestrian flow into the public plaza.

项目位于无锡盛岸地区，北临京杭大运河，南望太湖，地理位置优越。地块内地铁3号线、4号线交汇，占据丰富的交通资源。

以正圆占据地块，在场地内部形成围合，并在四个角落形成广场空间，将城市居民引入用地。圆形建筑也是对周边多样复杂的城市肌理的一种应对，激发积极的城市生活。

Chapter Four

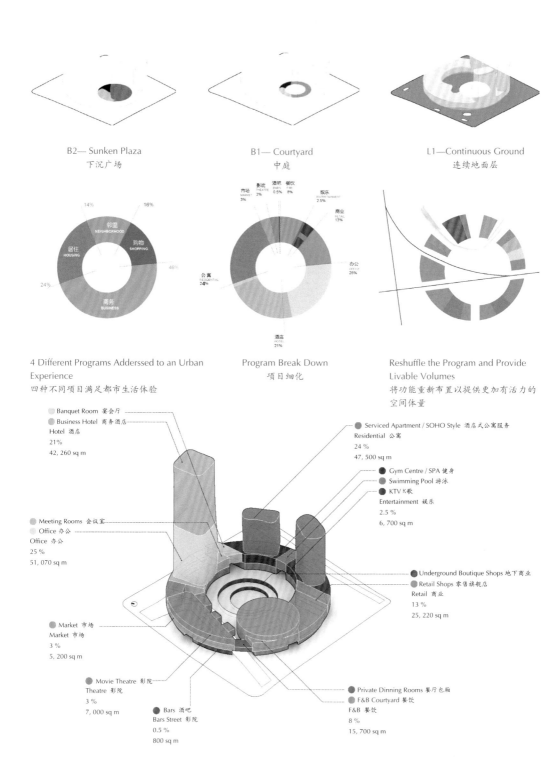

Wuxi Metro Mixed-use Project

L2—Protected Wooden Deck
受保护的木质平台

L3—Aerial Walkable Canopy
雨棚顶部的空中人行通道

The project is designated to become a vibrant city center, featuring offices, apartments, entertainment and commercial usage. Vigorous economical and sociological researches were conducted to create a fully-integrated megablock that is efficient fully engaged at all times throughout the day.
This architecture is a re-interpretation of the Hakka Village. The circular courtyard serves as a lively communal square for the inhabitants of this space. Overlapping, stacked rings also form elevated platforms to provide new retail experiences.

项目定位为都市活力中心，功能以商业、办公、公寓、娱乐为主。通过功能的细分和综合，实现功能的全天候使用。利用圆的围合在内部生成院落，给予建筑生动的公共空间。将层叠的环形走廊设置成空中平台，打造立体的新式院落商业空间。

Chapter Four

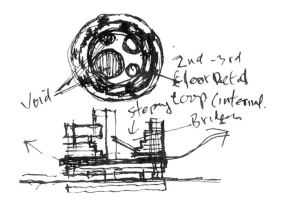

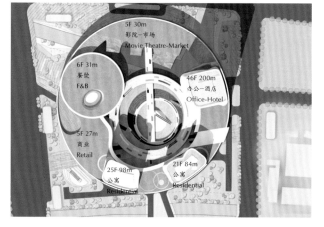

Ringed corridors wrap around the courtyard in various elevations, invoking the dynamic elements of the original Hakka architecture.

The sunken plaza morphs into the subway hub to provide seamless spatial continuity for the large flux of pedestrian traffic. A variety of transportation nodes are staggered along the sunken plaza, adding concise, yet energetic vectors of movements across the site. The architecture opens up to the city on ground level and forms aerial corridors at levels two and three, linking all flagship stores as one commercial network.

围绕中央庭院的环形走廊串联起不同标高的城市空间，重塑了土楼院落式的动感。

下沉式广场与地铁换乘枢纽无缝衔接，充分利用地铁这一巨大的人流源。沿此下沉广场，设多种竖向交通，形成简洁且极具趣味的动线。建筑首层向城市开放，并在二、三层设置环形空中走廊，贯穿各个商业旗舰店。

Wuxi Metro Mixed-use Project

Chapter Four

Ramps at the center of the courtyard connect the underground retail of plot A with three levels of cinema at plot B, creating a unique three-dimensional axis to emphasize the integration of programs within the community.

庭院中心设独立舒缓的坡道，连接A地块地下一层商业与B地块三层电影院，形成空中游憩景观步道，在地块内构成独特三维空间轴线。

Wuxi Subway Metro Project

Chapter Four

The project will stand as iconic, innovative vision for the city, one that embodies the circle as concept, greenscape as color, and courtyard as community spirit, all together enveloped by a contemporary facade.

以圆为魂、以绿为意、以院为心,项目立面设计上兼顾时代感、标志性、创新性,旨在塑造城市新形象,成就区域新亮点。

Wuxi Metro Mixed-use Project

Chapter Four

Gift Box   礼品盒

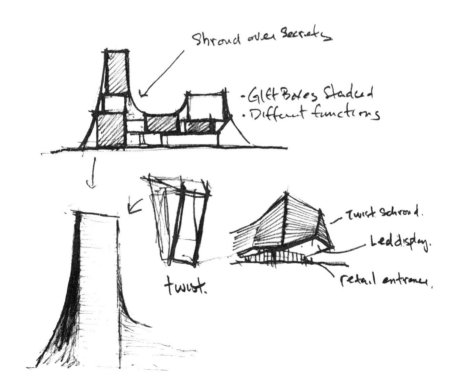

A spirited family gathers around the fireplace, singing, dancing and exchanging warm conversations. Colorful gift boxes are stowed neatly under the elaborately decorated tree, perfecting the familiar holiday scene.

There has been a long tradition of gifting between friends, families and acquaintances. The wrapping of the gift adds a shroud of mystery on top of the excitement of anticipating a gift. The opening of a gift box can generate an element of surprise, one of the most positive emotions among the innumerable human sentiments.

This project is a large electronics flagship store and retail headquarters. Electronics and household appliances allow us to escape from miscellaneous daily chores and physical labors, thus providing our lives endless choices and possibilities. We selected "Gift Box" as the project's starting point to convey the message of "packaging a colorful life." The project is not a mere commercial center; it is also a tribute to contemporary living.

精心装饰的常青树下，挂着堆着色彩缤纷的礼品盒，家人围绕着唱歌跳舞，尽情欢乐。这样温馨的场景是否能触动心中最柔软的部分？

把礼物放在礼品盒中送给亲人朋友，这种方式有着悠久的传统。在打开礼品盒的一瞬间，能激发收礼者所谓"惊喜"的情绪，这是人类众多情绪中颇为正面积极的一个，也是人与人的交往中很令人愉悦的元素。

本案是一个大型电器零售商的旗舰店和总部大楼。我们的日常生活已被各种家用电器所包围，得以从繁杂琐碎的劳动中解脱出来，电器在很大程度上改变了我们的生活方式，为每个人提供了更多的选择性和可能性。我们选取"礼品盒"这个切入点，试图传达"打包多彩生活"这样一种信息。所呈现的不只是一个简单的城市商业体，更是一曲关于现代城市生活的欢歌。

Chapter Four

Beijing Suning Plaza

北京苏宁电器广场

Location: Beijing, China
Type: Mixed-use
GFA: 61,150m²

项目位置：中国 北京
项目类型：综合体
项目面积：61,150m²

Beijing Suning Plaza

Located within the Fourth Ring Road with a grand view of the CBD, Beijing Suning Plaza is a multi-functional complex that includes dining, entertainment, gyms, and SOHO-style offices. The goal for the project is to create an animated streetscape, to enhance lifestyle, and to promote commercial functions. The design incorporates a colorful, transparent box to express vibrant architectural programs and activities. The skin is composed of horizontal lines aimed to express the connection between architecture and site, creating a grand urban impression.

北京苏宁电器广场位于北京东四环，远眺CBD，位于CBD东扩范围内。作为一个集电器连锁、餐饮、娱乐、健身及SOHO办公等多种功能的城市综合体，积极打造活力街区，提升城市生活品质，体现品牌形象。设计所采用的彩色玻璃盒子意喻丰富的建筑功能和使用者在其中的多种活动。外层包裹的建筑表皮和横向的肌理旨在表达建筑与基地的紧密联系，同时塑造大气的城市形象。

Chapter Four

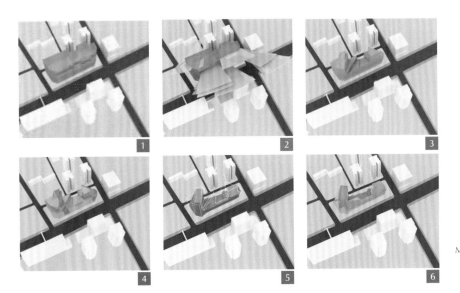

Massing Concept
体量概念图

The architectural massing is based on studies on site conditions and limitations such as:

1. The 80m height limitation.
2. Based on the orientation of the sun, the massing needs to be divided to create appropriate forms, especially for the northern residential units.
3. Ensure that the tower occupies the western portion of the site; that the podium is higher in the south and lower in the north, and the east side is slightly elevated.
4. Designating the eastern side for commercial entrance after an elevated plaza space; the center of the site is for an outdoor commercial plaza.
5. Twisting the floor plates to maintain the original irregular architectural form.
6. Unique textures on the facade express a fresh and grand appearance.

建筑体量的确定，是基于对地块各种限制和地理条件所进行的全面分析。

1．建筑限高80米。
2．依据大寒日北京的太阳照面和北侧住宅的采光窗位置，对建筑体量进行切割，反推出可满足日照要求的建筑形量。
3．要求塔楼只能占据场地西侧，裙房由此将呈现南高北低的形态，东侧可局部高起。
4．东侧作为主要商业入口，通过悬挑形成入口广场空间。同时在基地中部首层局部架空，形成室外商业广场。
5．通过楼板的扭转统一原先不规则的建筑形体。
6．肌理独特的表皮包裹建筑，使建筑形象既新颖又大气。

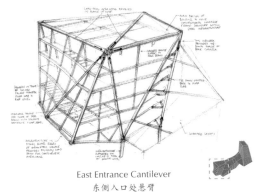

East Entrance Cantilever
东侧入口处悬臂

As its basic framework, the tower uses a tube-like structure that begins at the base of the podium, which will provide structural support for the cantilever at the eastern entrance, the partial-interior commercial plaza, and spaces with large structural spans.

塔楼采用框筒结构，裙房部分为框架结构。东侧入口处的大悬臂，建筑中部半室外商业广场上的大跨度结构以及塔楼主体的扭转都可通过局部钢结构实现。

Beijing Suning Plaza

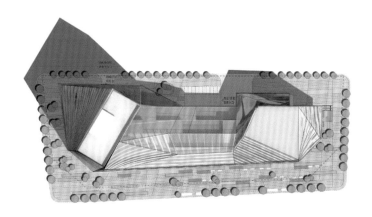

The northern podium is created by a series of steps that include dining functions, forming a rich and dynamic recreational space. The staggered steps on the facade incorporate colorful glasses reflecting different functions within the building. With consideration to Beijing's weather, the patio is covered with an extensive glass ceiling system. The glass wraps around the building's northern side echoes its south facade and enhances the patio function. The interior incorporates green design to minimize the green house effect.

北侧裙房设置层层退台，与餐饮功能相结合，形成丰富的休闲空间。退台立面采用彩色玻璃，使空间显得活泼轻松，与其功能相符。同时考虑到北京的气候，露台之上覆盖大的玻璃顶棚。采用玻璃包裹的方式既与建筑南立面的处里相呼应，同时也增加了露台的使用效率。结合室内的绿化设计，还能产生立体温室的空间效果。

Chapter Four

Beijing Suning Plaza

The curtain wall implements a 1.2m linear spacing. The facade incorporates an optical illusion to increase the building's mass. The intersections of louvers are utilized to enhance the overall appearance and provide sustainable functions. A portion of the podium incorporates a gradual transition in color to revitalize the commercial atmosphere.

幕墙采用1.2米间隔横向构件，呈现切片的形式，利用视觉错觉，增加建筑的体量感和稳定性。穿插的竖向百叶窗既能丰富立面层次，也可起到建筑节能的作用。裙房部分利用色彩变化，活跃商业气氛。

Chapter Four

Ink Painting 水墨画

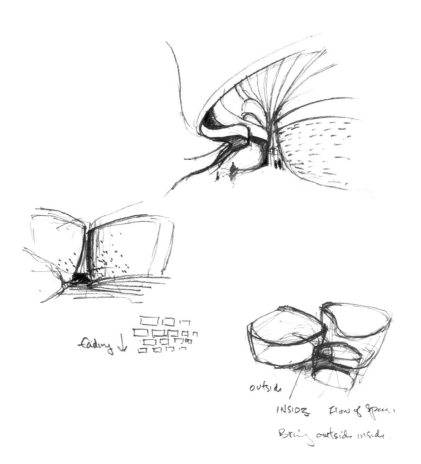

Beijing Artist Village Gallery Project

Pure, clean and simple, the ink painting is one of China's oldest arts. Black and white brush strokes can achieve a variety of shades, generating spectacular, magical scenes of the world. Ink painting is also called "color ink"; a name that reflects the change in shades of color. In a gradient of inked intensities, colors are born into the colorless. The limited canvas is thus expanded into an infinite realm.

The ink paintings embody a unique Chinese view of the universe. Located in Beijing, the project was commissioned for an older gentleman with an astonishing collection of ink paintings. We seek to replicate the marvelous aesthetics of the paintings in an architectural language.

Elements of both the "real" and the "imagined" are manifested as spaces in architecture, and as shades of ink paintings. Referencing the ink painting composition, we broke up the typical museum spatial boundaries. Traditionally non-virtual, or real space boundaries, are blurred by a gentle curve marked by different densities of separated objects, in a delicate but ample intensity. Even though the materials used were mere traditional grey bricks and white walls, we render the building a canvas for unlimited thinking and imagination.

In architectural spaces, we hope to provide an opportunity to identify "the meaning beyond the words, the flavor beyond the taste."

水墨画是中国传统绘画的一种，纯用水墨作画，黑白两色之间，通过用笔墨浓淡幻成世间万象，气象万千。水墨画有"墨彩"之说，只需用墨的浓淡变化即可体现色的层次，无色中生出有色，虚实中生出气韵，有限画布上生出无限境界。

水墨画所体现的是中国人独特的宇宙观和审美趣味。本案位于北京，业主是一位酷爱水墨画收藏的老先生，他的藏品给了我们最初的震撼。久久回味，如何用建筑的语言，来重现画带给我们的美好体验？这正是打动我们，让我们全心投入的原动力。

"虚"与"实"是建筑与绘画的通用元素，于建筑指的是空间意义上的，于绘画指的是浓淡意义上的。借鉴水墨画的构图，在美术馆中我们打破空间的界限，传统意义上非虚即实的空间在这个项目中被柔和的曲线和不同密度的分隔物所模糊，层次更为丰富，变化更为细腻。即使在材料上只使用了传统的灰砖和白墙，整个建筑也能引发无限的思考和想象。

我们尝试提供的是一次有"言外之意，味外之旨"，令人回味的空间体验。

Chapter Four

Beijing Artist Village Gallery
北京世纪国际艺术城美术馆

Location: Beijing, China
Type: Culture
GFA: 10,202m²

项目位置：中国 北京
项目类型：文化设施
项目面积：10,202m²

Award:
World Architecture Festival 2012, Winner of Future Projects / Culture Category

Chapter Four

Sited near a quiet countryside in Beijing, the Artist Village Gallery consists of a private museum and a private house. The inspiration of the architecture originates from a Chinese ink painting that blurs architectural boundaries with natural surroundings, resulting in a series of fluid spaces that provide the visitor a unique experience. This project attempts to express an architecture that captures the beauty of Chinese ink paintings.

北京世纪国际艺术城美术馆坐落于北京近郊，基地周边环境宜人，幽静自然。项目包含私人美术馆和馆长别墅两部分。设计灵感来源于馆长的水墨画收藏，模糊的建筑边界可与自然环境融合互动，流动的室内空间提供给参观者别致的观展体验。在这次设计实践中，我们尝试创造一种富有中国画意味的建筑空间。

Beijing Artist Village Gallery

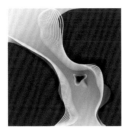

Spacial Concept
空间概念

Chapter Four

Entrance
入口

Beijing Artist Village Gallery

Chapter four

Beijing Artist Village Gallery

223

Chapter Four

Building Envelope Design Developmen

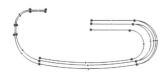

Plan Curves:
Offseted Arcs

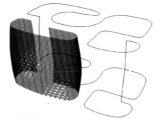

A Height Field Target Surface Is Created to Generate 4 Differnct Sizes of Windows According to the Relative Difference Between Target and Base Surface

Dome Design Development

Original Dome Surface

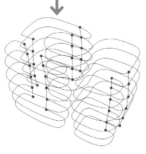

Control Curves:
Curves in Plan and
in Elevation

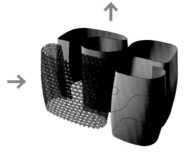

Diagonal Punctures Created in a Regular Module

Dome Is Pannelized into
Triangulated Flat Pieces

Panel Size Are Optimized to
Derive More Percentage of
Reasonable Panel Sized

Primary Structural System

Horizontal Band Added for Window Openings

Secondary Structural System

Brick and Window Implemented onto
the Secondary System

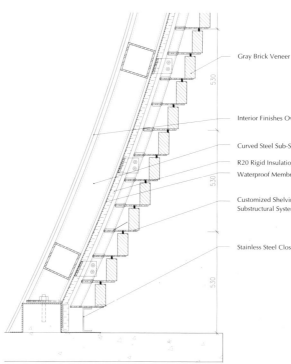

- Gray Brick Veneer
- Interior Finishes Over Air/Vapor Barriers
- Curved Steel Sub-Structural Framing System
- R20 Rigid Insulation 50mm Min.
- Waterproof Membrane Over Exterior Sheathing
- Customized Shelving System Bolded to Substructural System
- Stainless Steel Closure Gap at Base

Section Detail
墙身大样图

Beijing Artist Village Gallery

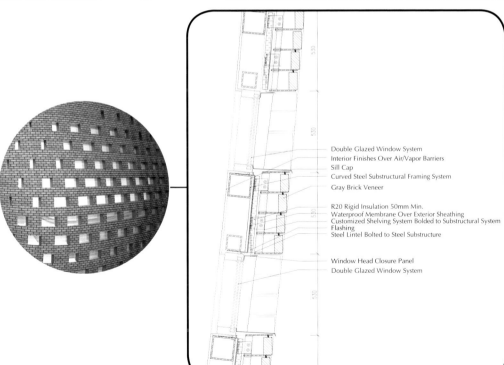

Double Glazed Window System
Interior Finishes Over Air/Vapor Barriers
Sill Cap
Curved Steel Substructural Framing System
Gray Brick Veneer

R20 Rigid Insulation 50mm Min.
Waterproof Membrane Over Exterior Sheathing
Customized Shelving System Bolded to Substructural System
Flashing
Steel Lintel Bolted to Steel Substructure

Window Head Closure Panel
Double Glazed Window System

Section Detail
墙身大样图

Chapter Four

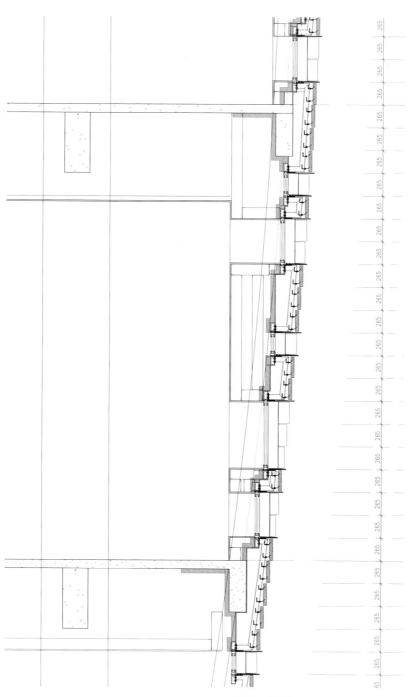

Section Detail
墙身大样图

Beijing Artist Village Gallery

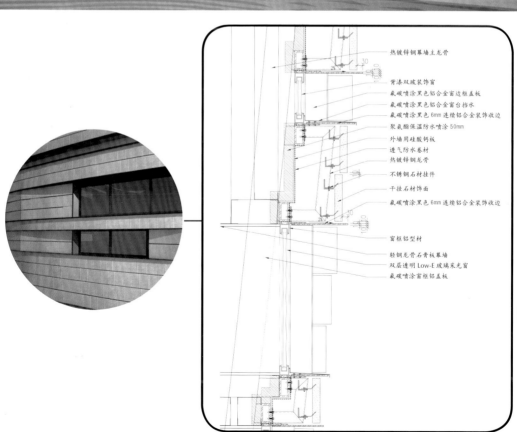

热镀锌钢幕墙主龙骨
背漆双玻装饰窗
氟碳喷涂黑色铝合金窗边框盖板
氟碳喷涂黑色铝合金窗台挡水
氟碳喷涂黑色6mm连续铝合金装饰收边
聚氨酯保温防水喷涂50mm
外墙用硅酸钙板
透气防水卷材
热镀锌钢龙骨
不锈钢石材挂件
干挂石材饰面
氟碳喷涂黑色6mm连续铝合金装饰收边
窗框铝型材
轻钢龙骨石膏板幕墙
双层透明Low-E玻璃采光窗
氟碳喷涂窗框铝盖板

Section Detail
墙身大样图

Chapter Four

The Kiss 吻

爱。超越战争,拥抱和平。

这是 Alfred Eisenstaedt 在纽约街头时代广场捕捉的瞬间。
正在庆祝"二战"结束的士兵情不自禁的亲吻美丽的女护士。

V-J Day in Times Square

Photoed by Alfred Eisenstaedt

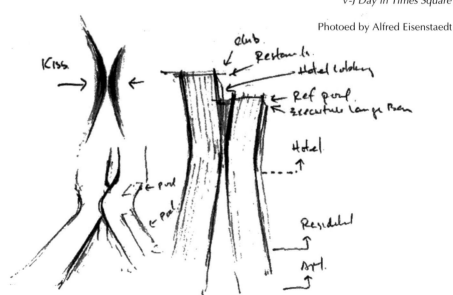

The project has taken the gesture of "kissing" as a design concept, emphasizing the romantic atmosphere of a coastal city environment. Formally, two towers incline toward one another. They are raised from the ground and joined together in a "kiss" at 160 meters in the air, forming an elegant high-rise tower. The two independent towers at the lower portion will house serviced apartments and residential units, the top portion after the two tower's reunion forms a 5-star hotel. The design also takes advantage of the height difference in its roof, creating a dynamic sky water garden space. The design principle is to maximize the views to the ocean; therefore the plan has taken an A-shaped layout to allow more than 90% ocean view guestrooms.

酒店设计利用"吻"这一概念凸显了滨海建筑的浪漫气息。在建筑形体设计上，利用两栋微微内倾的高层塔楼在160米的空中"相吻"，形成一栋完整的超高层建筑。上部融为一体的部分为五星级酒店，下部塔楼相对独立的两翼分别为酒店式公寓和住宅。利用建筑顶部的高低错落形成空中水景花园。裙房屋面延绵起伏，承托一大一小泳池并以叠水和浅滩相连，形成空间丰富的屋顶"水上公园"。

建筑形体有效降低风力对高层结构的影响，160米高处两翼相接的部分采用抗弯抗剪应力柔性结构，从而最大限度地降低了塔楼两翼位移对结构造成的不良影响。建筑设计原则为获得最大的海景观赏面，在塔楼平面的设计上采用A字型的布局方式，使海景房的面积达到90%以上。

Chapter Four

Haikou Changliu West Coast President Tower

海口长流西岸首府

Location: Haikou, China
Type: Mixed-use
GFA: 176,540m²

项目位置：中国 海口
项目类型：综合体
项目面积：176,540m²

This mixed-used project is located in Changliu, city of Haikou, Hainan Province. The site is at the intersection of Binhai Boulevard and Changbin Rd. overlooking the Wuyuan River Forest Park to the east, and the ocean to the north. The project's total construction area is 256,540m² including 176,540m² of above ground structure and 8,000m² below ground structure. The tower stands 250m tall, and contains five functional parts. The commercial podium starts at the ground level and allocates space for a Grand Ballroom, sky swimming pool and spa fitness, serviced apartments and residential units. The highest portion of the tower accommodates a 5-star hotel that contains 260 ocean view guestrooms.

本综合体项目位于海南省海口市长流起步区东北角35号用地，即滨海大道和长滨路交汇处，东临五源河森林公园，北临广阔的大海。项目总建筑面积256 540平方米，地上176 540平方米，地下8 000平方米。建筑高度250米。功能分为五部分，裙楼底层为商业，顶层为大宴会厅，空中泳池和SPA健身会馆、酒店式公寓和住宅以及五星级酒店设置在塔楼中。其中酒店海景客房260间。

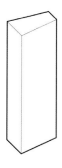 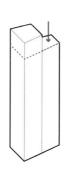 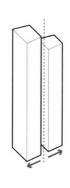 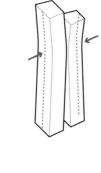 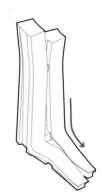

Massing Study
体量研究

Overall massing with optimized view orientation	Lowering the north end to create roof terrace	Seperating the massing to reduce the wind load	Kissing mass - Two towers are pushed towards the center	Extension of tower profile for the podium roof
可获得最大景观的形体	去掉北部顶部一角，获得屋顶平台	分成两部分形体，减小面宽过大问题，中间通透减小风阻	两侧向中心推进，形成优美的比例	一侧表皮向外伸展，形成裙房

Chapter Four

Skypool

High speed elevators take guests to the sky lobby at the 69th floor. Upon arrival, they are presented with the "skypool", an extension of the ocean view, through a vast glass curtain wall. A skybridge links the lobby to the opposite roof terrace, underneath the terrace there is an extravagant Executive Lounge.

天池

高速电梯将客人带至位于69层的酒店大堂，通过4层通高的落地玻璃，一汪水波与天色相连。客人通过室外连桥进入"天池"，连桥尽端有阶梯通向下方的行政酒廊，酒廊的玻璃顶将天光与粼粼水波引入室内。

The Kiss

The towers connect to one another at the 55th floor, allowing larger units of hotel rooms as well as floor to ceiling views of the ocean.

蜜月套房层

塔楼南北翼在55层相接，此处的酒店客房为跃层套型，并于中部设置高两层的落地观景玻璃朝向大海。

Podium Terrace Pools

The main pool and the children's pool connect via water; the podium terrace gradually raises to become the tower facade.

裙房屋顶泳池

主泳池与儿童戏水池通过叠水相连，商业裙楼屋顶连绵起伏，与塔楼底端自然衔接。

Haikou Changliu West Coast President Tower

Chapter Four

The Staircase 阶梯

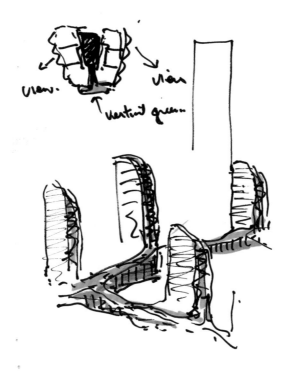

Keppel Land Chongqing Mixed-use Development Concept Masterplan

Chongqing is a city with stories. The romantic scenes of this mountain-city are used as background in numerous movies, which further add legendary color to the place. The Keppel Land project turns complicated site conditions into a distinctive residential and commercial complex, developing an urban living area expansion from traditional districts toward the Changjiang riverbank. Connected with the commercial corridor, which penetrates through the entire site from Kaixuan road to Renmin Park and connects the Jiefangbei CBD, a full commercial loop will be created upon the completion of the project. Residential blocks divided by retail streets will be reconnected by the landscape corridor, on top of which a harmonious community will emerge. The greatest feature of the residential towers are the vertical sky gardens. A raised podium landscape transforms into a vertical landscape, making the towers like the trunks of giant trees protruding towards the sun. Inspired by the Chongqing traditional stair street, stepped terraced landscape passages interweave into the pedestrian retail passages and connect multiple renovated historical buildings, creating a dramatic walking experience.

100 years ago, Marcel Duchamp investigated the possibility of continuing four-dimensional motion from a two-dimensional canvas in an ingenious manipulation of space and time. The radical concept, as showcased in his work, Nude Descending a Staircase No. 2, instigated a wave of vigorous re-evaluation of European social ideals, as well as a collective questioning and negation of the traditional artistic model. In the face of the rapid change of times, this constant evolution of interpretation and expression became the focus of contemporary thinking.

Stairs, as architectural elements, were quintessential in demonstrating Duchamp's revolutionary space-time concept. The stairs are viewed as a two-dimensional object in plan, yet already suggestive of movement, of ascension or descension in elevation. Ultimately, the application of time injects a four-dimensional motion towards the space above the stairs. The project is located in the mountain city of Chongqing, a site of undulating terrains, rich in architectural features, diverse in population and crowd circulation. Employing the concept of the stairs, a modern city composed of dynamic, continuous movements, emerge amongst the layers of this natural landscape.

重庆是一个有故事的地方。以山城为背景拍摄的大量影视作品赋予这个地方一种传奇般的色彩。吉宝置业凯旋路项目将复杂的场地条件转换为独具特色的住宅与商业综合体。方案将行人及城市生活从老城区引向长江水岸。通过建立始于凯旋路，止于人民公园，贯穿整个场地的商业步行走廊，可以形成连接解放碑商圈的完整商业环路。被商业街分割的各居住楼通过景观通道连接为和谐的社区。高层住宅以连续的垂直绿化平台为主要特色，并与裙楼屋面绿化贯通，仿佛参天巨树的主干。受到重庆传统楼梯街的启发，步行商业街顺应地势自然变化，串联多处历史建筑，创造一种有戏剧感的游历体验。

一百年前，杜尚 (Marcel Duchamp) 在画作《下楼梯的裸女》(Nude Descending a Staircase No.2) 中研究如何在二维的画布上展示四维时空的连续运动过程，作品的产生背景是20世纪初整个欧洲的社会反思和对传统艺术模式的集体否定。如今，面对又一个快速变革的时代，这种对周遭事物的阐释和表达的不断革新再次成为焦点。

阶梯也许是演绎杜尚时空概念的最佳场所，平地上物体的运动是二维＋时间，而在楼梯之上才是真正的三维＋时间的四维时空运动。设计中对楼梯概念的演绎充分利用了项目多变的原始地形、丰富的建筑功能和多样的使用人群，一个动态连续的现代城市综合体便在这自然山水中层层展现。

Chapter Four

Keppel Land Chongqing Mixed-use Development Concept Masterplan

吉宝置业重庆凯旋路项目概念规划

Location: Chongqing, China
Type: Mixed-use
GFA: 317,225m²

项目位置：中国 重庆
项目类型：综合体
项目面积：317,225m²

Keppel Land Chongqing Mixed-use Development Concept Masterplan

Keppel Land Chongqing Project sits on the river front site of the central district of Chongqing. The masterplan proposal contains program of residence, serviced apartments, retail and SOHO, also with a great potentials on the reuse of historical building as commercial attraction.

吉宝置业凯旋路项目位于重庆市中心滨江地带。项目功能包含住宅、酒店式公寓、零售和SOHO办公楼，同时需要对旧建筑进行再利用。

The Project is going to introduce new landscapes into Chongqing's commercial environment. Targeting the group of people who value lifestyle with premium quality, it is going to provide the most contemporary commercial experience.

此项目将为重庆商业环境增添新的景观。以城市中追求高品质生活方式的人群为目标，提供最具潮流精神的消费指向。

Chapter Four

The design of commercial spaces intends to inherit the unique heritage of Chongqing, especially the distinctive "Stair Street" in Chuqimen Area. Integrated with the planned flagship stores of prestigious international brands, the design creates a flexible, rich and dynamic shopping location while keeping an intimate feel.

商业空间的设计力图继承重庆传统，尤其是储奇门地区独特的楼梯街空间形态。结合国际品牌商铺规划，形成灵活丰富、动静结合的消费场所。

Differentiated from the normal shopping mall, the main features of this project include a series of open plazas and block-like commercial spaces. Platforms created at various heights in different open spaces of the site will provide a unique environment for designer brands and restaurants.

The renovated historical buildings are proposed to be the focal point on the retail streets, echoing the success of Keppel Land's Bugis Junction in Singapore.

区别于常见的各类大型封闭式商场，吉宝凯旋路项目以露天街区式的商业空间为主要特征。通过场地高差不同的公共空间而设置的平台，为特色品牌和餐饮提供别具一格的环境。

经过精心改造的历史建筑将成为商业街中的亮点，延续吉宝新加坡Bugis Junction项目中的成功经验。

Chapter Four

Original Site
原始地形

Gradient
梯度

Gradients and Roads
梯度与道路

Terraces White
台阶

Pedestrian
人行道

Commercial
商业

Landscape Path
景观道

Tower Development
塔楼

Keppel Land Chongqing Mixed-use Development Concept Masterplan

 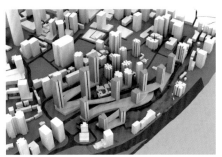

In order to offer a maximum diversity of unit types while keeping compliance with the rigorous regulations of Chongqing government, two kinds of residential buildings were created: slab and tower. Both types of building have sky gardens and shared yards so that the residents will have direct access to nature inside the city. Sky gardens would also become a main architectural feature being displayed proudly on its elevation.

住宅建筑分为点式塔楼和板楼两种，以满足重庆严格的间距控制规范，并提供户型的多样性。此两种楼型都配置有空中花园及共享庭院，以使居民获得最直接的自然体验，空中花园因此成为建筑立面形象的主要特色。

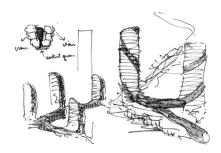 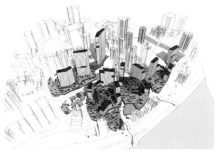

Two open corridors penetrate through the entire site and there is a landscape belt which connects three main areas between residential towers. Through effective property management, the landscape belt could supply residents with various activities, while at the same time provide some open spaces for the community.

尽管整个基地由两条开放的公共走廊所贯穿，住宅塔楼之间依然形成一条连接三个主要地块的社区景观带。通过有效的物业管理，可以使此景观带既满足住户的活动要求，又提供市民休闲的空间。

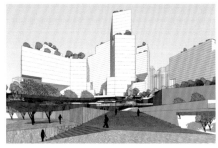 

The wide selection of different unit types (ranging from small serviced apartments to luxury duplexes) will meet the various needs of people with different backgrounds, so as to ensure the diversity of the community and enable the creation of a harmonious and vibrant living environment.

通过配置多样的户型（从小巧的酒店式公寓客房到豪华的跃层豪宅）来满足不同阶层城市居民的需求，保证社区内人口成分的多样性，形成和谐多元的居住社区。

Chapter Four

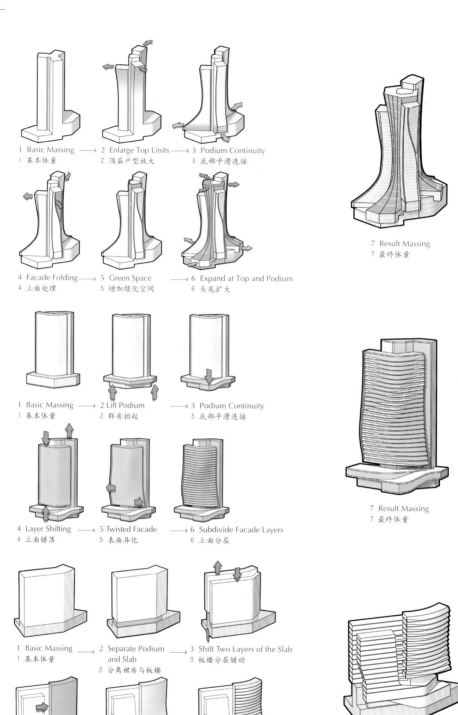

Chapter Four

Keppel Land Chongqing Mixed-use Development Concept Masterplan

Chapter Four

Memory 记忆

正是由于失去了与过去的联系,正是由于失掉了"根"——这种情形才造成了人们对文明的种种"不满",造成了这样的慌慌忙忙——我们才不是生活在现在而是生活在未来,生活在未来那黄金时代虚无缥缈的许诺里。

——《荣格自传:回忆·梦·思考》 荣格

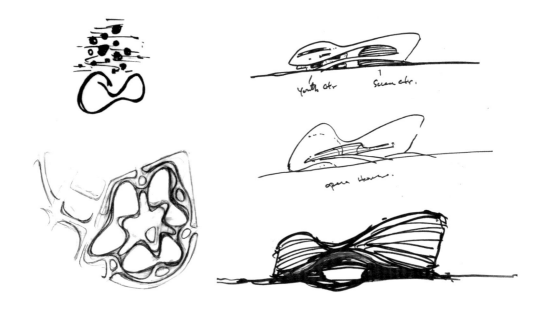

Guangzhou Nansha District Jiaomen River Central Area Southern Waterfront

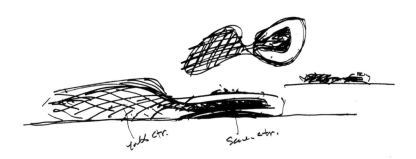

"Memories", is a very sweet word. Happiness is often the warm aftermath of a recollection, as pain is often the process of healing from a past, but also accompanied by the joy of survival. Time is the distance between reality and memory, a distance from where we can observe these moments time after time.

Memories are often distorted perceptions of reality, from which we conceive the notion of freedom by projecting our consciousness under different lights. What is a house? Is it a cubical container? Is it a definite manifestation of a certain object or space? What role does the occupant play? How can we render architecture into something fluid and nostalgic, something which serves as triggers to certain memories? Can the house stir hidden emotions, uncover hidden thoughts? Should we not have a more intimate relationship with the space we inhabit?

Cultural buildings could be seen as a materialized expression of memories related to a story, a period of time or even a single person.

回忆，是一个很甜美的词。快乐会因为过去式在一遍遍的描摹中越发温暖，痛苦会因为过去式被慢慢抚平，甚至伴随劫后余生的狂喜。时间为我们留下了一段从现实到回忆的距离，这段距离带给站在时间这一头的我们一种观看的美感。

回忆是对曾经的事实的变形，我们可以在其中找到某种自由，投射出自身，不同的个体对同一件事情可以有迥然不同的回忆。那房子呢？仅仅是四方的盒子吗？一个确定无疑的物体或空间？那使用者在其中扮演怎样的角色？如果建筑变得如同回忆般柔软，与过去的一些美好相联系，我们是不是可以在其中找到更多隐秘的感动？我们与一个房子的关系是不是可以更为亲密？

文化建筑常常可以成为回忆的载体，关于一个故事，一段时光，一个人。

Chapter Four

Guangzhou Nansha District
Jiaomen River Central Area Southern Waterfront
广州南沙新区蕉门河中心南部滨水角

Location : Guangzhou, China　　项目位置:中国 广州
Type: Civic & Cultural　　　　　项目类型:文化设施
GFA: 217,810 m²　　　　　　　项目面积:217,810m²

The "Nansha Origin" project is located at the Guangzhou Nansha New District, the southern part of the Jiaomen River. The land is planned to fulfill five programs: art gallery, museum, science museum, youth center and grand theatre, all in direct dialogue with the surrounding cultural mix-use complex, to become the source of culture, history and ecology.

"南沙源"项目位于广州南沙新区蕉门河中心南部滨水角,用地规划为美术馆、博物馆、科学馆、青少年宫和大剧院五大功能,旨在打造功能集约、辐射周边的城市文化综合体,成为南沙新区的文化之源、记忆之源、活力之源、生态之源。

Guangzhou Nansha District Jiaomen River Central Area Southern Waterfront

Chapter Four

Project planning and design extend beyond conventional practice. The aggregated layout combines five buildings into three clusters, enclosing landscape and water in the center. Two connected buildings become hosts for science, art and music. The architectural cluster simulates a sense of enclosure while forming a vibrant public space.

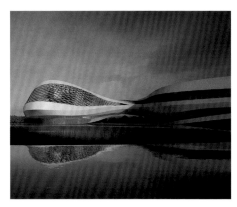

项目规划设计突破常规做法，采用聚合式的布局，"虚实相生"，将五栋建筑合为三组，围合中央景观水面，两两相连的建筑分别形成科学之门、艺术之门与音乐之门，营造建筑组群的场所感，形成充满活力的公共空间。

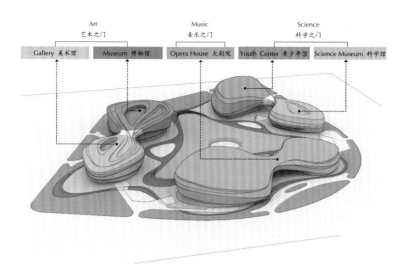

Function Analysis
功能分析

The architecture itself is inspired by fishing boats, fishing nets, plaited bamboo articles and other commonly seen tools in traditional local water life, and expressed through modern architectural interpretations.

建筑单体的灵感来源于渔船、渔网、竹编等常见的岭南传统水乡生活工具，通过现代手法，转化为不同层级上的建筑表达。

Chapter Four

Gallery + Museum

美术馆＋博物馆

Step 01
步骤一 —— 功能体量
Overall functional massing of museum and gallery
博物馆，美术功能体量

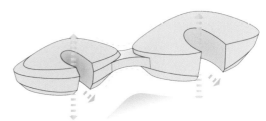

Step 02
步骤二 —— 吹拔空间和连接
Carving void in each floor to create patio space facing the landscape. Bridging two pieces of massing to create shared space
建筑临湖面临预留通高吹拔空间。体量之间由空中连桥连接形成共享空间

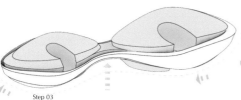

Step 03
步骤三 —— 包裹和提升
Skins wraps massing around to create functional exhibition space. Life the skin in between to form a "gate"
表皮包裹建筑体量形成展览空间。提升位于体量之间的表皮形成"门"空间

Step 04
步骤四 —— 切片
Slicing the skin to form horizontal windows facing the landscape, carving louvers to provide skylight for patio space underneath
在建筑表皮上切开细缝形成面向景观的横向条窗，体量顶部开天窗为吹拔空间提供天光

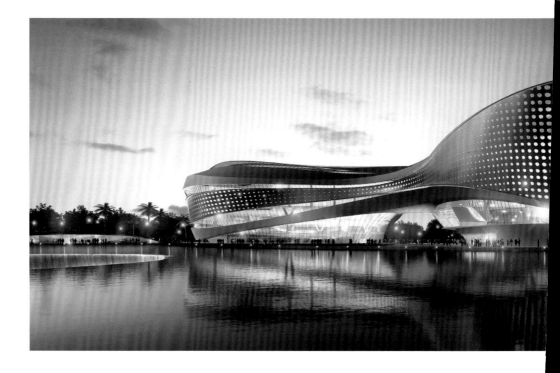

Guangzhou Nansha District Jiaomen River Central Area Southern Waterfront

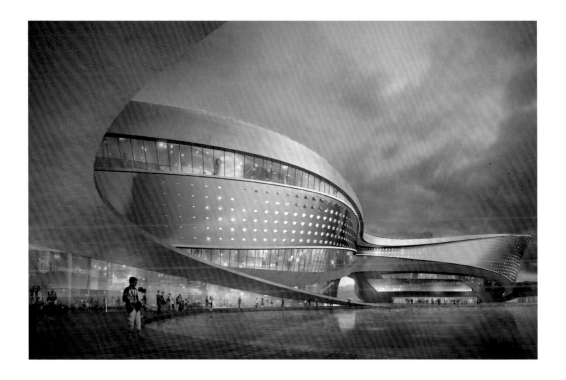

Chapter Four

Science Museum + Youth Center
科学馆＋青少年宫

Step 01
步骤一 —— 基本形体
Overall functional massing of youth center and science museum
青少年宫，科学馆基本功能体量

Step 02
步骤二 —— 纽带连接
Create 3d bridge to connect youth center and science museum. Visitors can pass through from one side to another
青少年宫与科学馆之间用连桥连接。游客可以经由连桥于两个建筑间穿行

Step 03
步骤三 —— 扭转缠绕
Twist a stripe of science museum's elevation into the roof of youth center
扭转科学馆立面体量形成青少年宫的屋顶

Step 04
步骤四 —— 细化雕琢
Emphasize the twisted strips to create net facade
强调扭转体量形成网状立面

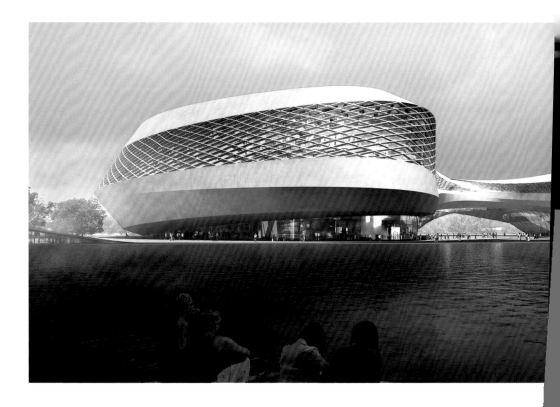

Guangzhou Nansha District Jiaomen River Central Area Southern Waterfront

Grand Theater

大剧院

Step 01
步骤一 —— 体量
Overall massing of the grand theater
基本功能体量

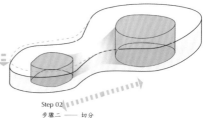

Step 02
步骤二 —— 切分
Split the opera and concert hall in two, create the public spaces in between
将歌剧院和音乐厅切分，创造两者之间的公共空间

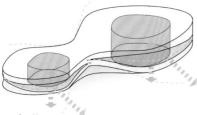

Step 03
步骤三 —— 切分，提升
Split the massing to create view vista at the atrium of both opera and concert hall. Lift ground floor in middle for pedestrian access
切分歌剧院和音乐厅体量为其中庭创造景窗。提升体量之首层部分，形成步行入口

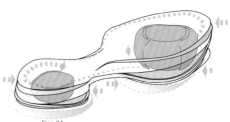

Step 04
步骤四 —— 衔接，剪切
Blend facade with hump roof to create continous change.Cut horizontal gaps to form strip window and ground level entrance
利用山脉状屋顶衔接两个建筑形成连续变化的体量。立面上剪切横向裂缝形成条窗和首层入口

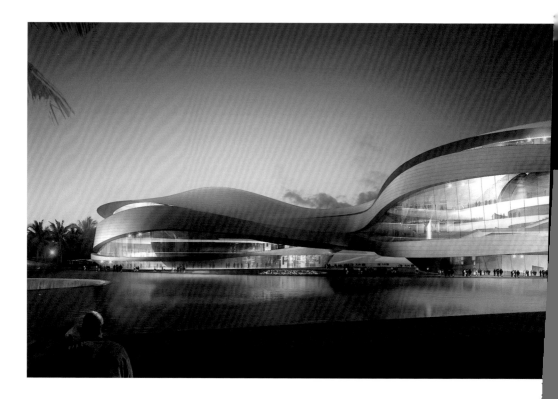

Guangzhou Nansha District Jiaomen River Central Area Southern Waterfront

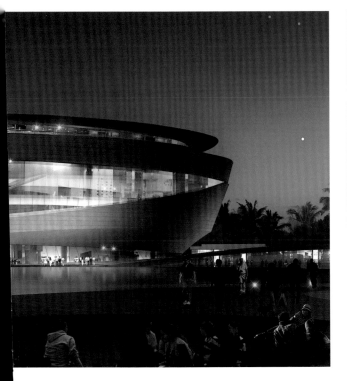

CHAPTER FIVE 第五章

For the earth, air is the most essential element and the origin of life.

For an architect, architectural creation is closely influenced by his formative experience and social background, just like the air around the earth, these influences have reached many aspects of his architectural development.

空气弥漫在整个世界,生命因其而生。

对建筑师而言,建筑创作也不是无端产生的,个人的成长经历和社会背景都会深刻地影响创作。这些因素犹如空气一般,建筑师置身其中,呼吸成长。

Background of Andy

Andy Wen was born in Taiwan. At a very young age, he migrated with family to the United States, where he later received his Master of Architecture degree from the University of Pennsylvania. In the 1990s, Andy went to China and obtained a Doctoral degree from Beijing Tsinghua University, he soon returned to Taiwan to teach in Ming Chuan University and later became the Head of Architecture Department. Andy continued to develop his design philosophy in the teaching process. Through a few renovation and expansion projects of campus buildings, he further refined his design theories in the practice.

In order to practice his own design theory at both a larger scale and wider range, Andy began to participate in the design of commercial projects for an international architectural design firm. Benefiting from years of academic experience and accumulation of theory, he was able to project a more comprehensive understanding of projects, fresh perspectives to the design, and ultimately, found breakthrough solutions to a design challenge. Meanwhile, his courage in pursuing bold and innovative design ideas and his interactive open working method brought a collective architectural enthusiasm very different from other commercial firms.

Mr. Andy Wen and Aedas met in 2008, he is currently an Executive Design Director at the firm. Aedas has provided a comprehensive platform for Andy's professional design practice giving his team abundant support to meet various design challenges and to create excellent architecture.

Andy has been working as a Design Architect for over 20 years with rich working experience in Asia. He was involved in numerous projects including master planning, mixed-use developments, apartment complexes, villas, 5-star hotels, office towers, cultural and educational centers design. Working with international firms in the United States, Taiwan and China, Andy has developed an extensive client network. Over time Andy has worked with a variety of international clients ranging from real estate investors, hotel chains and international commercial corporations to smaller-scale hotel developers and residential property firms.

Apart from architecture, Andy has additional experience with interior and custom furniture design. He has created over 16 lines of designer furniturewith styles ranging from modern to Asian contemporary.

Meanwhile, Andy is also a part-time Professor at Tsinghua University and Central Academy of Fine Arts in Beijing, where he teaches graduate level architecture to design students. It is within this combined professional knowledge that Andy is able to examine architecture from different angles. In a sense this is intriguing and not only beneficial but essential to balancing creative enthusiasm and self-reflection. Andy maintains close friendships with many contemporary artists and is influenced by their ways of interpreting the world. Perhaps it is this positive interaction with his surroundings and endless curiosity that has cultivated such a steady stream of creativity.

This book includes many projects that bear witness to the amazing creativeness of a mature architect in practice with a sophisticated design office.

Chapter Five

Andy and His Teachers(From Left to Right): Wu Huanjia（吴焕加）, Andy Wen（温子先）, Guan Zhaoye（关肇邺）, 2004

温子先先生的经历

温子先出生于台湾，幼时移居美国，在宾夕法尼亚大学获得建筑学硕士学位。20世纪90年代回到中国大陆，在清华大学获得博士学位。之后回到台湾，就任铭传大学建筑系主任一职。在执教过程中，温先生进一步对自己的设计思想进行完善，并通过多个校园改造扩建项目，对建筑理论进行了实践和修正。

进入新世纪，温先生为了在更大的范围和尺度上实践自己的设计理念，开始了在国际建筑设计事务所的从业经历。得益于多年的学术积累和思考，他总能对项目进行更综合全面的解读，带给设计团队多层次的启发，拓展设计思路，找到解决问题的突破点。于此同时，他在设计实践中敢于大胆创新，进而结合其互动开放的工作方式，总能为设计团队带来更多的热情和活力。

温先生与凯达环球建筑设计咨询公司的合作始于2008年，现任凯达环球建筑设计咨询(北京)有限公司设计执行董事一职。凯达环球为温子先先生的设计实践提供了一个成熟的商业和技术平台，专业的综合设计团队在温先生的带领下，承接了众多不同类型的项目，优秀的建筑作品也相继诞生。

温先生从事建筑设计逾二十年，拥有丰富的项目经验，曾参与多种不同类型的项目设计，包括总体规划、高层住宅、公寓楼、别墅、五星级酒店、办公楼等。温先生先后在美国、中国台湾和中国大陆工作，至今拥有庞大的客户网络，客户中包括地产投资商、连锁酒店集团、国际商业机构、小型酒店发展商和住宅物业公司等。

除了擅长各类型的建筑设计外，温子先对室内设计也十分在行，温子先至今已拥有十六个系列的原创家具，作品呈现出多样的设计风格。

作为清华大学及中央工艺美术学院的客座教授，温先生目前还在从事教学工作，任教建筑设计研究生课程。正是这种双重的职业身份，使温先生可以从多个角度来审视自己的建筑创作，有利于保持对创作的热情和自省能力。同时，温先生也与很多当代艺术家保持了亲密的朋友关系，从他们对现实世界的解读和表达方式中受益良多。也许正是这种与周遭的积极互动，对新生事物的无限好奇，孕育了温先生源源不断的创造力。

这本书中所收录的项目，见证了一位成熟的建筑师与一家成熟的设计事务所结合所迸发出的惊人创意。

Social Background of Andy's Design

In mainland China Andy Wen is involved in many significant construction projects. In this new world of architecture, Andy and other designers are facing the challenges and opportunities of a complex environment.

With an integrated Chinese and Western background Andy Wen holds a cultural advantage which allows him to practice finding a balance between the "unfamiliar" and "familiar." This advantage allows one to find aspects of particular sensitivity that can lead to breakthrough in an otherwise traditional heritage.

The conflict between the speed of development and design quality is unavoidable. While the world's shockingly fast paced development provides a great stage for the architect, the compressed schedule also gives the designer tremendous pressure. Many decisions are made in a hurry resulting in a collecting number of crude projects that exist throughout our cities today.

Andy Wen has perfectionist tendencies in the design practice. To create good design in a fast paced market Andy Wen utilizes his experience along with the professionalism of the team to avoid making mistakes or giving up on design quality. Despite these practices he sometimes has to make controlled compromises based on firm principles.

Speeding up the development of impulsiveness has been the collective responsibility of society. The impact of globalization on local culture is helplessly facilitated by invisible forces. There is no time to transfer thinking from individuals to groups to counteract these influences. This is the process by which our kitsch consumer culture has crowded people's lives. This may be a necessary process for a developing country, but we still believe that changing the status quo of power should be rooted in profound traditional culture. Although it now might seem very frail, more and more people are beginning to reflect that a single spark can start a fire. As an international architect, Andy Wen has had tremendous consistency and dedication towards "representation" in his efforts to support the transformation of traces of Chinese architecture into this era.

Large-scale real estate developments present issues in both unification and repetition which need resolution by the architect. In the design process the owner is not the end user of the building which is the fundamental problem Chinese architectural design is facing. From the 1980's "Matchbox" to the present neo-classical renditions, a market-oriented country with a large population has become accustomed to a "herd" mentality that emphasizes this stereotype. The loss of personal voice limits the end user's ultimate selection but also deprives the right of the architect's creation. Andy Wen has attempted to introduce a new architectural vocabulary and image into his projects and to break the stifling situation. In some projects this positive attitude has achieved extraordinary results and received the support from developers and local governments. Perhaps, at this current stage architects must stand by their intentions and their own aesthetic standards. Architects must also support efforts to persuade and carry out multi-party communications to arrive at interesting solutions.

Needless to say Andy Wen sees China as the best platform to practice his design. Despite many challenges, architecture has always been called upon to solve complex and contradictory issues. This land has an extensive cultural history, the world's largest population, a unique political and economic system, a vision and the determination of national revival. Here one can nurture a "narrative Architecture ", because there is so much to say.

温子先先生设计创作的社会背景

温子先所参与了中国大陆的诸多建筑项目。与所有奋战于这片热土上的建筑师一样，他所面临的是机遇与挑战并存的复杂局面。

中西结合的教育背景给了他某种文化优势，可以在"陌生"与"熟悉"之间找到平衡点，使他同时在传承与突破之间具有特别的敏感度。

发展速度和设计质量之间的矛盾是无法回避的。世界让人震惊的发展速度固然给了建筑师一个极大的舞台，但被极尽压缩的进度表也给了设计师极大的压力，很多决定都是在仓促之间完成，造成了粗制品的泛滥。温子先有某种完美主义的倾向，面对这样的矛盾，他利用自己的经验和团队的专业性来尽可能避免决策的错误和设计质量的降低。有时不得不作出某些妥协，但这种妥协也是有原则和控制的。

超速发展造成整个社会的浮躁情绪蔓延，全球化对本土文化的冲击造成集体的茫然，只能无奈地被无形的力量牵着走，从个体到群体都来不及思考。在这个过程中，媚俗文化、消费文化充斥人们的生活。这也许是一个发展中国家必经的过程，但我们依然相信改变现状的力量是扎根于深厚的传统文化中的。尽管，它现在看来如此微弱，但随着越来越多的人开始反思，星星之火可以燎原。作为一位具有国际背景的建筑师，温子先也从未放弃在项目中"再现"中国本土文化的尝试和努力。

大规模的房地产开发所造成的单一性和重复性也是建筑师所要解决的问题。设计过程中的甲方并不是最终建筑的使用者，这是中国的建筑设计所面临的根本问题。从20世纪80年代的"火柴盒子"到现在的遍地新古典主义，都是以市场为导向的产物，多人口国家惯有的"从众"心理，更加剧了这种千篇一律。个人话语权的丧失，在剥夺使用者的选择权的同时，也剥夺了建筑师的创造权。温先生尝试在他的项目中引入一些新的建筑语汇和形象，以打破这种沉闷，这种积极的尝试在有些项目上取得了很好的效果，也得到了开发商和当地政府的支持。也许，在现阶段，建筑师能做的是坚持自己的审美标准，并努力去进行多方的说服和沟通，以找到一个折中的解决方案。

温先生把中国视为实践他设计思想的最好的舞台。尽管面对诸多挑战，但建筑从来就是要解决各种复杂和矛盾的问题的。这片土地有悠久的文化，有全世界最多的人口，有独特的政治经济体制，有民族复兴的宏愿和决心，这里可以孕育出"会说话的建筑"，因为它有太多可以"叙说"的内容。

Social Background of Andy's Design

Postscript

Book of Fables
(Postscript)

Parables is an assemblage of experiences collected in recent years of architectural practice. Each project employs the idea of "reproducing the uncertain" as a driving force in design, actualized with the aid of modern engineering.

The building today is no longer a mere functional object. As an architect, I focus more on the presentation of a semantic exchange, a thorough and faithful interpretation of each project. To me, the most inspiring aspect of architectural design lies within the endless possibilities of form. Each project possesses a remarkable narrative, composed of unique circumstances for design regarding its particular environment and inhabitants. Thus, architectural design identifies and analyzes these unique existing conditions and possibilities of each site in the goal of deriving solutions specific for each project. It is a process that is both challenging and rewarding. To this day, I still maintain the same enthusiasm for each project with a high respect for the diverse possibilities of form. Admittedly, urbanization and globalization in a mass produced society are rapidly changing our surroundings in a way that renders many of our modern cities and structures monotonous and devoid of life. I hope to give each building a distinctive voice. Of course, this voice is not necessarily one of exclamatory declaration, it can be also be one of quiet guidance or suggestive provocation. This voice does not belong to me as an architect, but to what existed previously to my creation as well as what became of it afterwards; it belongs to the process of creation itself and a collection of all alternative factors. What we have done here is opened a path to the hidden story to give people an opportunity to experience the unseen.

After *Parables*, there are still boundless unknown territories, waiting for our venture.

The Story Behind The Story

This collection of buildings can be read as a series of stories. I would like to have a few words about these stories.

Writers

Aedas is a true international design team. Different cultures, personalities and methodologies are united under a singular drive for architectural creation.
With Aedas, work and life are not always separate realms. Bits of life outside the office are likely to trigger certain moments of brilliance in design, as the fruits harvested from work can also add a dash of color to life outside the office. The architect receives stimulation from all life experiences. Inspiration can stem from unexpected places.
Our team has grown to understand the negotiation between persistence and compromise. While we seek to express our individual voices, we also strive to create a positive work environment enabling the exchange of ideas and unity towards a common goal. A strong "orientation towards the people" is a concept all architects are familiar with. The product of architectural design is a vessel and habitat for the users of the space. As a result of design, the affects on the physical and mental condition of "the people" is an important element has too often been neglected.
I have always paid close attention to the unique background of each project, which includes its history, geography, as well as the cooperation of its designers and the implementation of the concept of "spaces made by people, for people". In China, the speed of the building process as a whole is executed through the collective operation of the government, developer, designers and consultants. These operations are sometimes mature and efficient, sometimes requiring time and considerations for outside factors. All of the above provide interesting challenges and result in diverse achievements.

Writing The Story

The process of architectural design is typically defined in four phases: conceptual design, schematic design, design development and construction documents. As an international firm, Aedas often collaborates with local design institutes in specific regions of the world in order to receive the appropriate qualifications to fulfill the specific requirements for construction documents.

Concept Design

This phase aims to compose a storyline for the project. The concept is not a mere product of an arbitrary idea; it is also formed by a series of preliminary research, case studies and communication between the clients and relevant consultants. After the requirements and facts have been gathered and clarified as outlines, the team will "brainstorm" to finalize the design. In the process of "brainstorming", all members are invited to give individual input, exchange and discuss ideas and choose a direction that will lead to the most rational solution.
The product of the "brainstorm" is not achieved in a linear manner. The team will take multiple approaches into consideration. Often, numerous conceptual developments will be presented for the exchange of comments, critiques and suggestions between designers, clients and consultants. The conclusion to these discussions will narrow down the result to one to three design options.

Schematic Design

The end of the conceptual phase signals the basic solidification of the general form of the spaces, as well as its functions. The schematic phase is crucial to the success of the project. We must remain loyal to the design concept while taking many other factors into consideration, such as the efficiency of the space, the quality of construction, and the social and economic impact of this operation. These logistic means must be fully integrated into the final design.
This phase requires a team composed of fresh-minded creative with strong comprehensive understanding in 3D, experienced architects with years of practice, as well as project managers capable of efficient resource allocations and excellent communication. A balanced team paves an excellent foundation to the success of the project.

Design Development

This is a transitional stage between schematic design and the production of construction drawings. This phase will address numerous architectural details, including the structural, mechanical and electrical systems, as well as lighting, landscape, interiors, building logo and cost control. On top of the design team, a large group of various consultants in these specific areas will also be fully involved at this stage. Architects will assume the role of the leader in expanding the design, coordinating with these consultants and mediating between concept and construction. This stage is often misunderstood as a strictly technical phase dictated by the engineers. Here, the architect is responsible in ensuring that the original concept and intent of the design carried out consistently through all phases of the project.
Building materials are also largely chosen at this stage. These choices are imperative, for this directly impacts the final outcome of the presentation. If the budget and schedule permit, suppliers and construction companies provide 1:1 scaled sample units to confirm the architect's comparative judgment in material selection.

Postscript

Construction Documents (Details Design)

For projects in Mainland China, this phase is usually completed by the local design institute, while the architect is responsible for making decisions on design details and auditing the final construction drawings.
Chinese cities have become a playground for international designers. Numerous foreign offices have claimed their part in the shaping the architectural grounds in China. However, not many of these works possess the well-founded quality to withstand scrutiny, much due to the frequent neglect to details and the indifference of the construction industry to employ long-lasting materials, methods and technology. This is an unavoidable problem with contemporary architecture in China. However, with the increase in building quantity, local design institutes and the construction industry are accumulating more and more experiences, the level of finish to new projects is certainly improving. As long as we are able to learn from our experiences from past international projects, at the same time factor in the status of China's current construction industry and produce plausible design solutions, we can increase the degree of completion, raise the standard of construction to produce longer-lasting buildings.

Rewriting The Story

The revision is a regular part of daily work for the architect. This operation is often repetitive and predictable in nature. It has become inevitable in any project, as complete smooth sailing is extremely rare in design. How does the architect deal with the week, month or even year of work being discarded at once? How does the architect find the motivation to re-start a design from point blank? Can the cycle of endless revisions be prevented?
Adequate, holistic project management from the initial stage is conducive in avoiding unnecessary design changes in the future. For example, the University of Liverpool came very close as a physical manifestation of the original concept conceived by the design team. This was all made possible through vigorous preparation, research and efficient project management.
Some changes are unavoidable due to external influences, such as when the client changes programmatic requirements, or when government officials don't agree with a certain style of design. Under these circumstances, one can come to appreciate the chance for modifications as a means of telling the "story" from a different approach.

Misinterpreting The Story

In many cases, a text will be "misread". In the past, "misreading" was considered a "wrongful" act, or a sacrilegious misinterpretation of the original "truth" as intended by the writer. In the modern era, "misreading" has become an art in its own. For many contemporary literary masters, these various interpretations to a text can be ambiguous, refreshing, paradoxical or even playful.
I have long practiced the design philosophy of "consolidated re-presentation of uncertainty", which expresses a progressive advocacy towards "misinterpretations". In the case of architecture, spatial and temporal qualities can be enriched through the implementation of such "uncertainty". The "consolidation" of various factors is a basis for this "uncertainty". The building is not a response to a singular factor, but a derivation of many different factors, and sometimes, this derivation can be a result of "misinterpretation". The "re-presentation" is an alternate means of depiction, not a one-dimensional repetition of the original. During process of translating the story using different "dialects", in what ways is the audience informed of the original content? This method of expression itself would naturally produce an "uncertainty."
"Uncertainty" implies disparity, distinctiveness and unpredictability. To exist in the fullest extent, one must not neglect emotions and thoughts that fall outside the realm of everyday life. This existence is a portrait of the

Postscript

designer; it is a display of remarkable individual freedom. There is no one "true" interpretation for the stories written in this book. We encourage you to experience these stories repeatedly, from different perspectives and arrive at different conclusions.

Ultimately, the building relies on its own voice to narrate the story. We welcome you to come along and experience the story.

寓言书后
（后记）

寓言书集结了我近几年的建筑实践，通过一个个项目，探索将"综合再现的不确定性"运用于实际工程的可行性和所能展现出的推动力。

现今的建筑早已不是单纯的功能性器物，作为一个建筑师，我所专注于呈现的更多的是一种语义上的交流和对每一个项目的真诚解读。于我而言，建筑创作最大的吸引力来自于个体的独特性，每一个项目都有唯一的"天""地""人"的背景，而找到那个最终属于这个项目的"天时、地利、人和"的故事则充满了乐趣和挑战性。至今，我仍然可以保持对每一个项目的热忱，不满足于简单重复自己的设计风格，这正是出于对多样性的尊重。不可否认，城市化和全球化正在迅速改变我们的周遭，旅行于不同的城市，不断被全然没有差异性的无趣感冲击着，我更希望可以让每一个建筑发出自己的声音，独一无二的声音。当然，这样的声音并不一定是宣言式的，可以是浅吟低唱的，抑或质朴无华的。这声音不属于作为建筑师的我，而是作品诞生之前、诞生过程中和诞生之后的所有因素的集合。我们所做的，是给这些隐秘的故事一个渠道，让人们有机会去发现。

寓言书后，依然有无数个精彩的未知领地等待着。

故事的故事

某种意义而言，您所读完的这本建筑作品集可以被看作一本故事集，在这样一个集子的最后我想聊一聊关于这些故事的故事。

写故事的人

凯达环球拥有真正意义上的国际化设计团队，不同的文化背景和各异的性格特质的人，是共同的职业理想将他们汇聚在一起。

成为凯达的建筑师，工作与生活便无法全然割裂，生活中的点滴都有可能激发设计中的灵感，工作中的收获也能给予生活新的色彩，无法想象一个对生活不敏感的人如何能成为一个好的建筑师。我们的团队有作为建筑师的坚持，也有不得已的妥协，这既是个体的状态，也是群体的境况，如何令团队保持一个积极的心态是公司和我所一直努力的。"以人为本"的理念大家都不陌生，而我们尝试把"人"的概念从项目使用者扩展到项目实施者，他们的身心状况也许是长期被忽略的一个重要问题。

我一直都关注每个项目独特的环境背景，这里所指的不单是历史地理意义上的，也包括各合作和执行团队所组成的"人"的环境。在中国速度和中国环境下，政府、业主、设计师、各技术顾问作为项目的执行主体各司其职。不同项目面临不同的合作模式，有时成熟高效，有时需要较长时间的磨合，但我们把这些都看作成就项目差异性多样性的积极因素。

我们这样写故事

通常一个项目的建筑设计部分可以分为概念设计，方案设计，扩初设计和施工图设计四个阶段。由于凯达是国际建筑设计事务所，在某些对施工图资质有特殊要求的国家，需要具有相应资质的当地建筑设计院配合完成施工图设计阶段。

这个阶段旨在完成一个故事的构思。大部分项目的构思并非依靠灵光一现，而是基于扎实的前期调研、案例分析、与业主和相关策划顾问公司的沟通等。经过理性的准备后，设计团队会用"头脑风暴"的方式最终确定设计概念。在"头脑风暴"中，团队中的所有成员都会在讨论板上贴出自己对项目的理解和想法，形式不限，大家在没有拘束的氛围中畅所欲言，最终合议出较为合理的深化方向。

"头脑风暴"的成果通常不会是单方向的，经过一定的整理会产生几个立足于不同思考角度和侧重点的初步概念方案。我们与业主和各顾问公司进行中期成果沟通，最后集结各方建议和意见，产生一至三个最终的概念方案。

方案设计

经过此阶段，项目的形象、空间、功能得以基本落实成形。这个阶段对项目的成败至关重要，既要忠于最初的设计灵感，又要充分考量功能的合理性，施工质量的可控性，建成运营的社会效应，几乎所有的影响因素都应该在此阶段被有效地整合，以确保项目的顺利推进。
这个阶段中，设计团队中既需要富于创造力和三维思考能力的设计师，也需要有多年实践经验的建筑师，还需要具备优秀沟通和资源配置能力的项目管理人员，一个均衡全面的团队是完成一个优秀项目的基础。

扩初设计

此阶段是方案与施工图之间的一个过渡阶段，包括结构、机电、幕墙、灯光、景观、室内、标识、成本控制等的相关顾问团队会在此阶段全面介入，构成一个大型的项目设计团队。建筑师在这个阶段中需要承担团队领袖和协调者的责任，把每个设计顾问的内容汇总到建筑图纸之中。对项目的整体把控在扩初阶段显得尤为重要，这个阶段常被误解为一个较为技术性的阶段，其实项目的各个阶段都应该不忘初衷，分工细化后由谁来确保项目的整体性、一贯性、建筑师责无旁贷。
建筑材料的选择通常也在这个阶段完成，这对最终作品呈现的实际效果无疑是相当重要的，必须极为慎重。如果预算和工期允许，一般会要求供应商或施工单位在现场制作1:1的样板，以便建筑师比较、判断、确认。

施工图设计（细部设计）

对于中国大陆的项目，此阶段一般是由地方设计院配合完成的。我们主要负责建筑细部的设计以及最终施工图纸的审核。
中国城市已成为国际设计师的游乐场，无数国际设计中国建造的庞然大物矗立在城市之中，但真正经得起推敲的精品并不多，理念上的束缚是一方面，细部设计和施工工艺的落实也是不可回避的问题。随着工程量的增加，无论是地方设计院还是施工团队都积累了越来越多的成熟经验，项目的完成度正在不断改善。只要我们能运用国际项目的经验，同时考虑中国现有的施工能力，设计出合理可行的细部节点，项目的完成度和持久性都是可以保证的。

故事的重写

方案修改是建筑师日常工作的一部分，这种调整有时是颠覆性的有时是常规性的，过程中的进进退退是不可避免的，一帆风顺的项目极为少见。如何面对一周、一个月甚至一年的工作被推翻？如何调整情绪重新开始？是否能尽量避免颠覆性的修改？
科学的项目管理会在项目的最初就对影响因素有较为全面的考量，设计概念的贴切与突出，都有利于避免不必要的设计修改。以西交利物浦大学项目为例，由于前期扎实的准备工作以及设计过程中的专业化管理，最终完工的成果与最初的概念方案极为相近，基本实现了团队的设计想法。
有些修改是由于无法避免的外部原因，比如业主对项目定位及功能的调整，政府对项目风格的不同意见等。对于这样的修改，不妨看作一个新的开始，一次新的机会，去重新寻找一个能打动自己的故事。

故事的误读

文字在很多时候都会被误读，曾经"误读"被认为是负面的，但在现代，"误读"成为很多当代文学大师所热衷的游戏，模棱两可，似是而非，絮絮叨叨。
我长久以来实践的设计理念——"综合再现的不确定性"中包含了对"误读"的开放姿态。建筑有别于文字，某种意义而言它所包含的时空因素更为丰富，所呈现出的语义的"不确定性"更为广泛。"综合"是这种"不确定性"的一个基础，建筑不是对单一因素的回应，而是所有因素结出的一个果，有时甚至是无意识的，正是这个果具有了模糊的形。"再现"是一种表达方式，不是单纯的临摹复读，而是一种传译，用不同的"方言"来讲述，听的人能收获怎样的信息？这种表达方式本身便具有"不确定性"。
"不确定性"意味着思想的个体性，生存的实在感，不是冷漠的于己无关的思辨，而是纯然属我的倾情。它是设计师意志的隐形，赋予"个体"最大的自由，这本书中的所有故事都没有标准版本的解读，如何理解完全取决于作为读者的您。我们仅期待对您而言这是一本有意思的书，是可以反复翻看而得到不同体会的书。
当然，建筑终究应该靠其本身来发出声音，所以欢迎您来到建成项目的现场，亲自体验和倾听，但愿会是值得回味的经历。

Team
团队成员

WEN, Andy 温子先	HUO, Ning 霍宁	YUE, Feng 岳峰	WANG, Dongwei 王冬维	LIU, Yan 刘燕	FENG, Ying 冯莹	LIN, Zihuan 林子欢
LIANG, Kuo 梁阔	YANG, Nan 杨楠	QIAN, Yijun 钱逸筠	YU, Qingbo 于清波	YU, Qianqian 于倩倩	HAO, Shuang 郝爽	WANG, Xiang 王翔
LU, Ran 芦燃	ZHANG, Zhe 张哲	WANG, Yu 王禹	LIN, Zhihong 林智宏	GUO, Wenyuan 郭文渊	LIN, Kaiming 林凯铭	YANG, Yongze 杨永泽
LI, Yunlong 李云龙	WANG, Siyu 王思羽	GAO, Wen 高雯	XIA, Yang 夏杨			

Former Team
前团队成员

BAO, Rui 包瑞	CAO, Jingjing 曹晶晶	CHAN, Augustus 陈忠威	David Bates 大卫·贝茨	DUAN, Xiaorong 段小融	FU, Mingcheng 傅明程	Gabriel Ball 盖波尔·鲍尔
GUO, Wenfei 郭文斐	HU, Xingzhu 胡兴竹	Jihyeun Byeon 卞智铉	KIM, Ja Young 金慈英	LANG, Xi 郎希	LEE, Charles 李泊	LIN, Diana 林诗韵
LIU, Hui 刘辉	LIU, Peng 刘鹏	LIU, Yuanxiang 刘渊翔	MA, Jianning 马剑宁	MENG, Fanli 孟凡理	Michael Cummings 迈克尔·卡明斯	Michael Doerneman 迈克尔·道尔曼
MU, Tong 牟瞳	NAM, Ji Won 南智元	PU, Jieyu 蒲洁宇	PU, Shengyi 濮圣一	REN, Xiaowei 任晓伟	RUAN, Jing 阮竞	Shannon Russell 山农·拉塞尔
SI Yang 司扬	TANG, Qing 汤青	TAN, Her Shyan 陈和贤	Tri Nguyen 阮仁哲	WANG, Viktoria 王楚羽	WANG, Zheng 王争	XIN, Lili 辛黎黎
YANG, Xinyu 杨新钰	YAN, Kai 阎凯	ZHANG, Jie 张婕	ZHANG, Yanlei 张彦磊	ZHANG, Zhen 张震	ZHAO, Chunjie 赵春洁	ZHOU, Lin 周琳

Team

Dr. Ning Huo currently works as Senior Associate with Aeads. She participates in the planning/design and project management for mixed-use projects.

Key projects include:
Suzhou Hong Leong City Center, Taipei Nangang Office Tower, Cheng Ying Center, Xiamen Green Land Mixed-use Project, Wuxi Metro Mixed-use Project, Haikou Changliu West Coast President Tower, Guangzhou Nansha District Jiaomen River Central Area Southern Waterfront etc.

霍宁博士现任Aedas高级主任设计师，主要参与大型综合体项目的规划、建筑设计及项目管理。

重点项目包括：
苏州丰隆城市中心，江阴澄星广场，台北南港办公大楼，诚盈中心，厦门绿地综合体项目，无锡地铁综合体项目，海口长流西岸首府，广州南沙新区蕉门河中心南部滨水角等。

Mr. Feng Yue has gained his Master of Architecture from Tsinghua University and Master in Urban Design from Harvard University. As associate in Aedas Beijing office, he has been responsible for several large-scale mixed-use projects, including 333 Shunjiang Road Mixed-use Project, Hengqin International Financial Center, COFCO Shanghai Joy City Phase 2.

岳峰先生拥有清华大学建筑硕士和美国哈佛大学城市设计硕士文凭。作为Aedas北京办公室的主任设计师，负责过多个大型城市综合体项目，包括门里成都顺江路333号综合体项目，横琴国际金融中心大厦，上海大悦城项目二期。

Within 5 years working in Aedas, Mr. Dongwei Wang participated in many large scale projects architecture design and planning, mixed-use, and facade renovation design. Mr. Wang is also the project manager in the project of Xi'an Jiaotong-Liverpool University Administration and Information Building, Xuzhou Suning Plaza, Zhengzhou Dennis David Plaza, Jinan Jiefangge Shimao Plaza, and Suzhou Hong Leong City Center.

在Aedas任职五年中，王冬维先生参与过众多大型项目建筑与规划设计，综合体和外立面改造设计。在多个项目中担任项目经理职务，例如西交利物浦大学行政信息楼，徐州苏宁广场，郑州丹尼斯大卫城，济南解放阁世茂广场和苏州丰隆城市中心。

Aedas Global Offices
Aedas 办事处分布——
业务遍全球　专技多元化

Global Reach
环球业务

伦敦|London
+44 (0) 20 3764 5450
london@aedas.com

香港|Hong Kong
+852 2861 1728
hongkong@aedas.com

北京|Beijing
+86 (10) 8529 0200
beijing@aedas.com

上海|Shanghai
+86 (21) 6157 0100
shanghai@aedas.com

成都|Chengdu
+86 (28) 8444 1338
chengdu@aedas.com

澳门|Macau
+853 2875 5530
macau@aedas.com

新加坡|Singapore
+65 6734 4733
singapore@aedas.com

洛杉矶|Los Angeles
+1 (310) 821 4859
losangeles@aedas.com

西雅图|Seattle
+1 (206) 452 1230
seattle@aedas.com

迪拜|Dubai
+971 (4) 3557 233
dubai@aedas.com

阿布扎比|Abu Dhabi
+971 (2) 6359 887
abudhabi@aedas.com

新德里|New Delhi
+91 (11) 4291 0000
india@aedas.com

Aedas Global Offices and Diverse Expertise

About Aedas
关于 Aedas

Aedas is one of the world's leading global architecture and design practices.

The company is built on the belief that great design can only be delivered by people with a deep social and cultural understanding of the communities they are designing for.

Aedas global platform for creative excellence in design enables some of the world's most talented designers to plug into the latest information and delivery systems they need to produce truly world-class design solutions.

The practice's unique structure, global presence, and commitment to cutting edge R&D are testament to our desire to deliver design excellence to clients wherever they are in the world.

Aedas 是国际首屈一指的建筑设计事务所。

Aedas 一贯秉持的信念是：只有对所服务的当地社会与文化具有深刻了解的建筑师才能做出杰出的设计。这一信念是公司的立身之本。

Aedas 为优秀的设计创意提供全球性平台，使得世界上一些极具天赋的建筑师可以获得所需要的最新信息，并得到设计交付系统的支持，以此来实现真正的世界顶级设计解决方案。

无论业主地处何处，Aedas 的独特组织架构、全球布局以及对尖端研发的全力投入，无不体现着我们为客户提供卓越设计服务的承诺。

Aedas Directors
Aedas 董事

Aedas directors with special contributions to projects mentioned in the book:
Aedas 董事们为此书中的项目做出了重要贡献：

Keith Griffiths on Hengqin International Financial Center, COFCO Shanghai Joy City Phase 2 and Suzhou Hong Leong City Center project.
Tony Ang on Suzhou Hong Leong City Center project.
Ed Lam on Hengqin International Financial Center project.
Larry Wen on Xi'an Jiaotong-Liverpool University Administration and Information Building project.
Benjamin Chan on 333 Shunjiang Road Mixed-use project.
Peter Marshall on Wuxi Metro Mixed-use project.
Kevin Wang on Xuzhou Suning Plaza project.

Acknowledgements
致 谢

Beijing Century Culture Properties Co., Ltd.
北京世纪文创置业有限公司

Chengdu Menli Wangjiang Real Estate Co., Ltd.
成都门里望江置地有限公司

City Developments Limited
新加坡城市发展集团

COFCO Land Shanghai Xinlan Real Estate Development Co., Ltd.
中粮置地上海新兰房地产开发有限公司

Earnest Development & Construction Corporation
诚意开发股份有限公司

Expo Shanghai Group
上海世博发展（集团）有限公司

Greenland Group Shanghai Oriental Cambridge Property Development Co., Ltd.
绿地集团上海东方康桥房地产发展有限公司

Guangzhou Municipal Bureau of Urban Planning / Nansha
广州市规划局南沙开发区分局

Keppel Land China Limited
吉宝置业中国有限公司

Pierson Capital Consultancy (Beijing) Co., Ltd.
皮尔森投资咨询（北京）有限公司

Sichuan Yanhua-zhixin Industrial (Group) Co., Ltd.
四川炎华置信实业（集团）有限公司

Sinobo Land Co., Ltd.
中赫置地有限公司

Sino-Singapore Tianjin Eco-City Investment and Development Co., Ltd. (Pre-Incorporation)
中新天津生态城投资开发有限公司（筹）

Suning Real Estate Group Co., Ltd.
苏宁置业集团有限公司

Suzhou Industrial Park Education Development & Investment Company
苏州工业园区教育发展投资有限公司

Wuxi Metro Group Co., Ltd.
无锡地铁集团有限公司

Zhonghong Holding Hainan Risheng Investment Co., Ltd.
中弘控股海南日升投资有限公司

Zhuhai Shizimen Central Business District Development Holdings Co., Ltd.
珠海十字门中央商务区建设控股有限公司

Special Thanks

Special Thanks
特别感谢

Keith Griffiths for the leadership in Aedas and trust in key designers to push for more meaningful design achievements.
纪达夫先生——对凯达环球的领导及对设计师们的充分信任成就了更具深意的设计作品。

Andy Bromberg for being an inspirational figure with fantastic projects realized out of Aedas.
安德宝先生——作为一位极具灵感的设计师为凯达环球带来了极具创意的项目作品。

Ken Wai for the management in Aedas China offices.
韦业启先生——负责凯达环球大中华区分公司的领导和管理。

William Wong for connecting Aedas designers to wonderful clients.
黄威林先生——为凯达环球的设计师找到了十分优质的客户。

Aedas BJ Staff for the hard work in design and technical support to make our designs realized.
凯达环球北京公司的同事们——在设计及技术领域的支持和辛勤工作使我们的设计成为现实。

Professor Guan ZhaoYe for reminding the importance cultural identity play in architectural re-presentation.
关肇邺教授——教授了文化元素体现在建筑表现方面的重要性。

Professor Wu Huanjia for reminding the importance of understanding the role creative innovations play in the history of architecture.
吴焕加教授——教授了不断创新的理念在建筑历史长河中的重要性。

图书在版编目（CIP）数据

寓言书：建筑的秘密 = Parables : architecture with hidden secrets : 汉英对照 / 温子先编著.—
天津：天津大学出版社，2014.5
ISBN 978-7-5618-5067-1

Ⅰ.①寓… Ⅱ.①温… Ⅲ.①建筑设计—作品集—中国—现代 Ⅳ.①TU206

中国版本图书馆CIP数据核字（2014）第096556号

《MARK国际建筑设计》杂志社 策划
地　　址：北京市朝阳区望京西路48号金隅国际E座12A05室
电　　话：010-84775690/5790
邮　　箱：info@designgroupchina.com
责任编辑：郭　婷　叶　玮　秦达闻
装帧设计：温子先
封面设计：温子先
项目执行：周丽雅

Parables: Architecture with Hidden Secrets
寓言书：建筑的秘密
温子先 编著

出版发行：天津大学出版社
出 版 人：杨欢
地　　址：天津市卫津路92号天津大学内（邮编：300072）
电　　话：发行部：022-27403647
网　　址：publish.tju.edu.cn
印　　刷：北京信彩瑞禾印刷厂
经　　销：全国各地新华书店
开　　本：163mm X 240mm
印　　张：18
字　　数：286千
版　　次：2014年6月第1版
印　　次：2014年6月第1次
定　　价：68.00元

（本图书凡属印刷错误、装帧错误，可向发行部调换）

著作版权所有 · 违者必究
All Rights Reserved